No. 553
$7.95

ELECTRONICS SELF-TAUGHT WITH EXPERIMENTS & PROJECTS

By Jim Ashe

TAB BOOKS
Blue Ridge Summit, Pa. 17214

FIRST EDITION

FIRST PRINTING—JULY 1971

Printed in the United States
of America

Library of Congress Card Number: 78-162408

Preface

This book is about a remarkable old technology: electronics, with historical roots extending back at least to 600 BC. In 600 BC, Thales of Miletus, one of the Ancients of History (maybe we should call them the youngsters), puzzled over his observations of static electricity. Particularly, he wondered why something he called an **elektron** attracted small particles after it was rubbed. Even then men were using electricity: shocks from electric eels were thought to alleviate stiffening diseases such as rheumatism. Thales ignored, or didn't notice, examples of electric repulsion as well as attraction, and he concluded that when amber was rubbed, little filaments of a "sticky smell" extended from it. He thought when the amber cooled the filaments pulled back in, tugging small objects with them. And it was more than two thousand years before anybody substantially improved this thinking.

Quite recently, as history goes, men called "alchemists" were looking for a "philosopher's stone." This was thought to be a magic material that turned base metals into gold, a sort of chemical Midas touch. There could never actually be a philosopher's stone, we know, because gold is only made naturally inside old stars when they explode: Chemical energy cannot support the reactions that make gold. Of course, the alchemists never found their philosopher's stone, but as they learned more about nature their art grew into our modern sciences of inorganic and organic chemistry.

Yet, might the alchemist's faith in his philosopher's stone have some basis in fact? A few years ago, archeologists excavating ancient Arabian ruins uncovered small clay pots with corroded iron spikes set in asphalt covers. These look remarkably like electric batteries. Maybe one day that old alchemist's belief will be traced back to some secret Arabian group whose trade was **electroplating** base metals with a thin layer of real gold.

True electrical science appeared in 1752, authored by Ben Franklin. What he did was more important than you may think. Not only did he study the behavior of electricity and make important discoveries about it, originate the terms "positive" and "negative" as replacements for a variety of terms then used to describe electrical polarity, and generally set electrical science on its feet, but he **engineered** the lightning rod. People say he invented it, but that's wrong. Franklin proposed a specific system of grounded spikes on building roofs, which he designed following much observation of how lightning acted and careful thought about why it acted that way. He ignored a lot of free advice that his work was against God, nature, and the then prevailing general scheme of things. He went ahead and designed electric circuits to do a particular job, **and he said why they would do it.** Now, more than 200 years later, his explanations are still acceptable. That's engineering, and it establishes Ben Franklin as the world's first competent electrical engineer.

Electrical science really grew in the 1800s. Volta, Ampere, Coulomb, Henry, Faraday, Ohm, Gauss, Newton, Gilbert, Maxwell, and many others left their marks on the field—and we still use their names to label the key ideas they helped shape into useful form. We use these men's work today to predict what our circuits will do if we build them. We don't need to puzzle over haywire assemblies, seeking trial-and-error results hardly usable elsewhere. And that brings us near 1900, when the modern kinds of electronics began with Thompson's discovery of the electron. Johnstone Stoney named it, and in 1909, Millikan developed a remarkable experiment that measured the electric charge carried by electrons, tested one by one.

So this is a book about a remarkable old technology. It's so old it goes back to ancient times and has been growing as long as our known history, longer, probably. And yet it is one of today's most important—that is, useful—technologies. But why does that make it remarkable? Because unlike data processing, for instance, and most other modern technologies, there is room for amateurs in electronics. Amateurs in electronics can choose to do simple, interesting things mainly for relaxation and a change in pace. More ambitious amateurs can try to understand something meaningful about electronics, and to design and build their own circuits rather than simply copying published material. It is those amateurs this book is largely directed to. Finally, really serious amateurs can work at developing their skill and competence until they become, as very many have, professionals in their own right. I think these, too, will find a value in this book.

Jim Ashe

Contents

Chapter 1

The New Electronics

The man who knows electronics can travel almost anywhere, and upon arrival he is sure to find the familiar basic principles of electronics hard at work in offices, laboratories, industries, and homes nearby. Electronics has a tremendous practical value, maybe this is why it makes a challenging and interesting hobby. Electronics is accessible to amateurs because it retains its close relation to basic science, even though it has branched into nearly all fields of technology. Because of their close connection to the great basic ideas our technology uses, the ideas of electronics can be boiled down to a point where children or amateurs from far outside technology can use them and they'll still work. It just isn't true that all powerful ideas are obtuse and complicated. As you learn more about modern electronics, and perhaps examine some pro type books, you'll come to see how convenient this is for the hobbyist—and how remarkable.

INTRODUCING THE NEW ELECTRONICS

The new electronics, whether for professionals or amateurs, places great importance on ideas. People still can concentrate on craftsmanship, if they want to, working as amateur wiremen to build impressive electronic machinery or assembling kits into simple, useful devices. But the real adventure today is in ideas and the way they can be used to design and quickly assemble diverse and challenging new kinds of circuits. The new electronics in no way resembles a craft or trade; its outlook is distinctly professional.

In fact, the modern trend takes us even beyond that. The new electronics engineer or amateur looks very much, these days, like a systems specialist. Not so long ago an extremely versatile high quality amplifier called an "operational amplifier" was available for maybe $75 and up. It needed expensive related circuits and power supplies. Now, while writing this book, I picked up three op amps for $3.90; they run off ordinary transistor batteries. These are the 709C amplifiers discussed in Chapter 7, and today people use these complex 14-transistor amplifiers more freely than they once used the far more expensive vacuum tubes. This is only one example of a powerful trend that is de-emphasizing circuits and making systems thinking important.

How It Grew

While electronics is soundly based in our technological history, modern electronics pretty clearly began during World War II and has been growing rapidly ever since. World War II was only partly a war of armies; important battles were won or lost according to how well the contenders appreciated the value of technological ideas and systems. The Japanese had working radars aboard their battleships before Pearl Harbor, for instance, and the Germans, too, built radar systems. But it was in England, and later in America that the value of radar was really understood and accepted, so the new technology received the terrifically concentrated work needed to make the radar idea into a large number of efficient, effective radar systems. As a result, world history was perceptibly changed. We are still hoping it was changed for the better.

All this research, and it covered very much more than radar, changed men's thinking about technology. Creative thinking became more important. Before WW II, for instance, there were few basic electronic circuits and some fairly definite ways to use them. Telephone technology was an important part of what passed for electronics, and if you were anything like an inventive type you probably weren't in the communications field.

But during WW II men had to come up with quick answers to tough problems. Often there wasn't any clear way to even describe these problems, much less their answers; you had to pick off a large chunk and start several lines of work to discover what it really was and what to do about it. This concentrated work broke down men's mental barriers against getting things done; they brought out an old tool that is always there but too often set aside: "Anything goes, if it works." Engineers looked closely at their vacuum tubes, capacitors, resistors and other components to see what these were really like. Then sometimes they ignored manufacturer's recommendations as to what these items were "for," and the state of the art advanced rapidly.

Such engineering and research brought us the new electronics. Soon after WW II, Brattain, Bardeen, and Shockley developed the transistor in the Bell Telephone labs. At that time the infant computer industry was having trouble with the vacuum tube's inefficiency, slow response and poor reliability. The second generation of computers was almost as free of vacuum tubes as the first was of transistors, and large computers no longer threw off enough heat to warm large buildings in winter.

Once solid-state technology proved workable, there was lots of money for more research. Integrated circuits appeared soon after transistors—the idea preceded them, in fact. By 1960 the vacuum tube was obsolete and integrated circuits were beginning to push out transistors; the early 1970s are seeing the advent of LSICs, complete systems up to complete computers, built on single chips of silicon crystal. The trend that brought us modern electronics will continue, but probably faster than now. But whatever happens, one fact is here to stay. The "experimenter" is out of modern electronics until he sits down with some good books and magazines and learns what it is all about.

A SIMPLE SAMPLE: REMOTE TEMPERATURE READING

Temperature interests people, who react strongly to extremes. Therefore, temperature measurement has a practical value, depending jointly upon its cost, its meaning, and the difficulty of getting usable results. It's good to be able to measure temperature remotely—to discover how hot or how cold an attic gets, to monitor hot-air system performance, to check the temperature of a water line too near subfreezing conditions. If temperature were easier to measure, it would interest people even more. While others concern themselves with louder or more wideband hi-fi systems, let's think about temperature—and circuit design.

You can design your own circuits and systems, working without special training. They can be more powerful and versatile, far more challenging than the following example. The key point is this: to learn easily, start simple. Study how ideas relate to things or purposes and practice recalling, finding, or inventing the appropriate ideas. Later, if it comes to that, think about hardware. An example of something simple and yet useful, is the thermistor, which is a chunk of stuff that carriers electricity, but only moderately well. Most conducting materials work better as they become cooler (superconductivity is this property gone runaway). A hot copper wire is a measurably less efficient conductor than is the same wire at room temperature. But a thermistor works the other way: as it gets hotter it conducts better, and this change is sharply temperature dependent. To see how a thermistor's temperature behavior compares with a copper wire's, see Fig. 1-1.

The curve in Fig. 1-1 shows how the resistance of a practical sized length of copper wire depends upon temperature. The resistance of 50 feet of no. 30 AWG copper wire at room temperature (68 degrees F) is 5.16 ohms. Cooled to -4 degrees F its resistance goes to 5.003 ohms; at the water boiling point its resistance is 5.474 ohms. Lab test instruments can measure these resistance levels well enough for this to make a very good thermometer. And copper wire is cheap, but test gear that sharply distinguishes between 5.160 ohms and 5.227 ohms (68 and 99 degrees F) and even smaller differences is expensive. Also, if the temperature-measuring piece of wire is any distance from the test gear connecting lead resistance becomes important. That's where the thermistor fits in. Over the same temperature range a thermistor's resistance might fall from 900 ohms to 7 ohms. If the thermistor always behaves the same way (it does if not abused), then why not use this sharp dependence on temperature to measure temperature?

We begin with a trip to an electronics service store, where the parts racks yield a good-looking thermistor. The one used in Fig. 1-2 is a television replacement unit. With one good specimen in hand, we proceed to find out what its properties are. Ice with salt in it gives us a practical minimum of -4 degrees F. A pan of ice or snow slush supplies 32 degrees F; room temperature is about 68 to 75 degrees F. Body temperature, reached after two or three minutes in your armpit, is 98.6 degrees F, and boiling water is 212 degrees F. We plot these points; though few they are enough. Thermistors aren't adventurous—a smooth curve gives us a general picture for converting resistance readings to temperature. All this is a little free of real accuracy, but it goes fast and gets us started. Precise results have their place—in the finished work.

Now, is measuring a thermistor's resistance and reading the resulting temperature off a graph truly electronics? Some may think not, insisting there ought to be a circuit. So, we will engineer a circuit. First, we must know the electrical properties of our thermistor. These are described by the curve in Fig. 1-1. To measure this resistance, we will compare it with a known resistance, avoiding the inefficiency of a general-purpose test instrument working at a fixed purpose job. A nice choice for this is the circuit of Fig. 1-2. This is called a "wheatstone bridge." Its meter reads zero if both meter terminals are at the same voltage. This happens when the left-

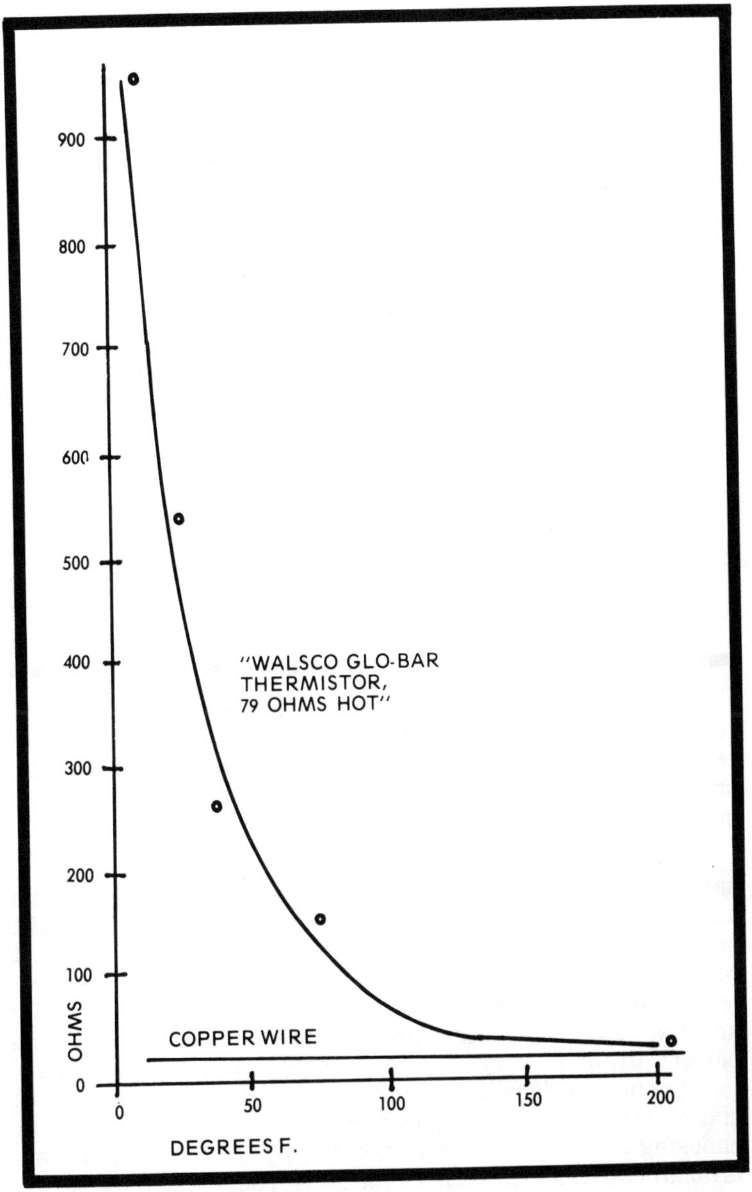

Fig. 1-1. The resistance change with temperature variations in a common service variety thermistor is tremendously larger than that of a piece of copper wire. This large variation is easily applied to temperature measurement.

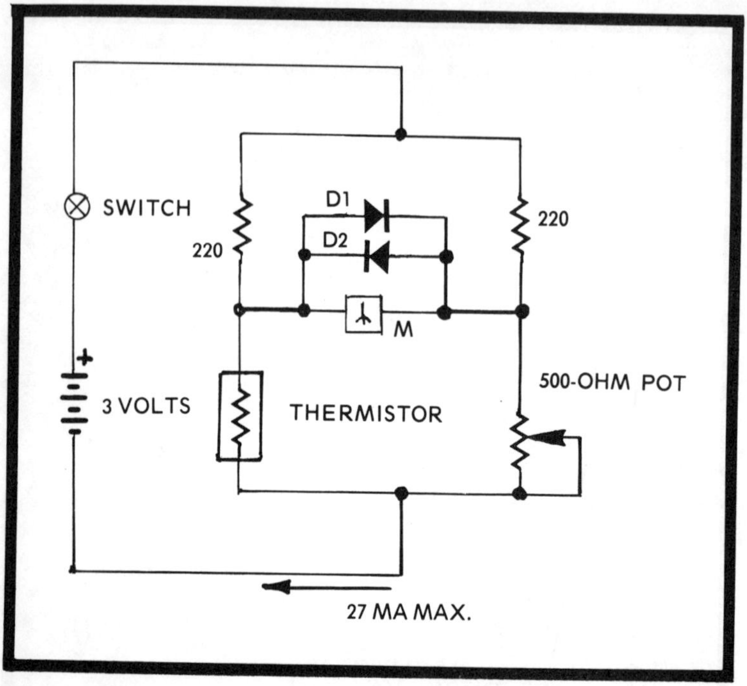

Fig. 1-2. This simple circuit measures thermistor resistance by comparing it with the resistance of a low-range potentiometer. Meter M can be a utility multimeter or a special unit. Diodes D1 and D2 can be any inexpensive germanium point-contact diodes.

hand voltage divider reduces battery voltage in the same ratio as the right-hand divider. What values might we choose for R1 and R3? If they are equal, the meter reads zero when the pot resistance equals the thermistor resistance. For best effecitveness their value should be in the same ballpark as R2 and R4, so 220 ohms looks good. This will result in tolerable battery drain at very high-temperature reading conditions.

As shown this will read down to only about 20 degrees F, since the pot is limited to 500 ohms. We could remedy this by choosing a smaller R3, a larger R4, by switching in an additional resistance in series with R4 (a range extender) or we might decide for now we aren't all that interested in temperatures under 20 degrees F. Two germanium diodes, one facing each direction, prevent excessive voltage from appearing across the meter, which is a multimeter for bench testing the idea or an inexpensive zero-center microammeter for a working instrument. Another try at the service parts store may turn up a suitable tuning indicator that will be quite happy in this new job, or a catalog item might look good. Notice that a switch extends battery life by hundreds of times, since now batteries deliver current only when checking temperature. And because this circuit compares resistances rather than measuring voltages, it isn't affected by battery aging. If it works, it works right.

When the overall circuit is assembled and working, why use the curve of Fig. 1-1 for temperature measurement? We aren't really interested in resistance; we want temperature readings. So we simply make a dial or scale-and-pointer, write numbers on it, and read temperature directly. Fig. 1-1 now helps by telling us where temperatures are located in between the calibration test values. Is that all there is to circuits and design? Hardly. Yet, this shows us some useful pointers about engineering. There's far less math than you might believe from checking some engineering books (which are stronger on analysis than design). By zeroing in on a practical case and using simple ideas simply, we avoid complexity, get along fast, and come up with a useful circuit.

ELECTRONICS FOR AMATEURS

Electronics technology has deep roots in science, history and everyday technology. It is particularly good for amateur work and study, since it is meaningful right now, and the student can put his newly learned ideas to work right now. This fact alone is responsible for terrific motivation and is an obvious aid to learning. The man who wants a change from his regular work, or who finds that his job lacks opportunity and meaning, can find exciting new outlooks in electronics. And a child interested in electronics stands squarely on a direct path to a rewarding and very useful career.

I think the realism of electronics helps make it even more interesting. You can try electronics from an amateur viewpoint, as a hobby, but there are no amateurs sitting in front of circuits on workbenches. **Everthing anybody does along this line is the real thing.** All electronic devices, circuits and systems work the same way for anybody, and if some people get more interesting, successful, or useful results than others, it must be that they understand the ideas better.

How far you go in electronics is up to you. Think about this, if you haven't already noticed it: learning things can be very hard work, but knowing things is then no work at all. And using your knowledge can be the most agreeable kind of experience, when it results in getting things done. Learning can be difficult, but is it really worse than wrestling with a circuit that will not function (probably because you have missed some key idea)? If you are short on money, there are lots of good books in the library, along with professionals, to help in your work. What you need to become a real expert is right there; in fact, I think a good amateur in electronics will progress a considerable way toward the pro's competence.

But there seems to be a considerable body of opinion against learning and study in amateur electronics, and I think I know what supports this. It is a reaction to too much bad writing in the field. Amateurs these days are regularly and routinely misinformed. The field is literally strewn with poorly written, inappropriate semitechnical literature turned out by authors who think technical writing for hobbyists is basically professional literature made sticky and with important facts and references cleared out. The pros cannot use that, and neither can anybody else. If you want to try something really needed, give a thought to writing technical material for nontechnical people. The market deserves attention, and the world needs this badly.

Engineering can be tough. Yet, in some ways, describing and designing circuits isn't all that hard. Often, somebody looks at basic laws such as Ohm's Law, or Thevenin's and Norton's theorems, or basic ways of using transistors, and he concludes "This must be made simpler, or else omitted." Omitting things is easier, of course, but the real-life circuit acts the same in any case and when the inadequately informed amateur tries to put the circuit into a new job, or has trouble with the original design, he goes down for a ten count. But these ideas are communicable, and mastering the hardest of them is really far less effort than many amateurs routinely spend trying to make things work. You can do better than that. As I see it, the really needed basics are here. The next few chapters cover them in fair detail. Then we will go into several projects that put them to work.

Chapter 2

The Private Electronics Lab

Your first do-it-yourself project should be your lab. You need a good place for more concentrated work. It will take something like two months to two years to assemble your lab, probably, unless you have lots of money for buying everything new immediately. Actually, you're better off without that money, since you'll have to think more. Consequently, you'll have a better lab and get more fun out of it, over the long run.

A word of warning! Look out for negative thinking. I have seen, time and again, some boy's father—or even a grown man—who says "This isn't serious; any old piece of junk will do till he (or I) learns better." This is a fallacy, something like that of the archer who saved money by practicing with crooked arrows until his accuracy improved to the point he could justify the greater cost of straight ones. He didn't realize his real investment was himself and his time. The archery world never did hear from him and years later his tackle was discovered, broken and dusty, in some unused corner.

If you are interested in electronics, spend enough time and money, and spend it well. This book is written for the man who means to do something, not for the negative thinker. The difference between using and spending your money is a subtle one you'll face often in electronics. It's another side of the similarity between saving money and investing it well. Try to put off buying anything until not having it hurts; then choose gear and parts you can use and reuse for triple utility from every dollar spent . . . I mean invested. Over a year or two, or three, this practice yields fantastic results.

FIND YOUR BEST SITE

First, where do you put your lab? Before you decide, think about what you will do in it. Modern electronics is a generally quiet and unsensational occupation; it resembles a game. There isn't a lot of machine work; low and high power audio systems can be tested into a load resistor rather than dumping tens of audio watts into the atmosphere. That pollution is avoidable. For the best time and money use, you'll read a lot, and I hope you will spend time designing circuits rather than making a continual copycat of yourself. What all this boils down to is that modern electronics comes nearer than many people think to being a living room hobby. In fact, the ideal site is just off the living room, unless you have hermit tendencies. In that case, maybe you can get over them. If you are driven into the cellar, the attic, or a garage, you probably have some preparatory work to do concerning electricity and climate control.

If you seek out some remodeling literature (try your local library before buying any books) you will be glad you did. The room needn't be very large and if it has quality type walls; the hazard to them is small. But if you have to put up something, I'd recommend half-inch plywood up to four feet, and quarter-inch above that. Careful shopping should turn up plywood at quite low prices, and it's much nicer to work with than sheet rock. Put all the thermal insulation you can find under the plywood, because it makes climate control easier. Remember the moisture barrier faces the inhabited side. Just plain varnish or even shellac over the wood is probably better than any other finish. Avoid time spent on color schemes. If you use paint I suggest a nice gray with a light ceiling. A cheery yellow looks good, too.

Electricity is discussed soon in more detail, but think in terms of surface wiring. You can change that easily. Grounding is less important than you may think; you want a vicinity grounding system, but not necessarily an earth connection. I'll come back to this later, too.

Now, climate control—that's important. If your lab winds up in the cellar you'll have a humidity problem. In that case, make necessary changes to limit air circulation and then bring in a dehumidifier. For one or two people in one room, you do not need a lot of air circulation, and a small dehumidifier can regulate air moisture admirably, if the turnover isn't too much. Water the dehumidifier takes from the air will have to be managed somehow; most models have a pail inside that you dump perhaps oftener than daily. Best time for dehumidifier buying is in the fall; you may be able to find one at below 25-percent of the list price. Remember to see it work.

In a cellar site you may do without an air conditioner, but one is good in other rooms in the house and in an attic you'll lose much time without one. But good thermal insulation plus a small air conditioner can manage almost any weather, and since the insulation is already there, maybe you can use electric heating in the winter. This takes lots of insulation, particularly if you cannot run up a really heavy electric cable, but it saves making expensive additions to your heating system.

The engineer knows modeling is cheaper and faster than making the real thing to see how it works out. You can model your lab various ways. Scale drawings are nice, if done lightly in pencil; inch-to-the-foot cardboard mockups also help you visualize clearly. Find a drafting or architecture book in your library for lots of good ideas about this. Whatever you plan, don't build for the ages. Build so it's nice to use, but easily changed. I guarantee you'll be making changes soon.

Maybe you'd be interested in the lab where this book's projects originated. Fig. 2-1 is a floor plan. It shows how the lab fits in a room corner, and there is lots of space left over when the chair and a couple boxes are tucked under the bench (where the servo voltage regulator resides). This setup with parts and storage shelf assembly on the left is appropriate for a southpaw; anybody else would set up a mirror image of this. One shelf, about bench height, has tools and work supplies laid out in the open; a toolbox is nice but it takes up space. Shelves at the top take the parts boxes described later; three lower shelves hold miscellaneous little used instruments (a luxury) and assorted finished projects. It is good planning to avoid collecting too much of this stuff.

Perhaps you think you don't know enough about what you will do in it to plan your lab. Plan anyway; planning and design are often uphill work and these paper operations really do catch bad mistakes, even if they do miss some, too. That's part of the challenge. Back in history Roger Bacon said "Truth will sooner out from error than from ignorance." If you have a plan, you can improve it; if you haven't, you can throw down your money in a grand gesture, and stand there as it whistles off in time's chilly breeze.

BUILD YOUR WORKBENCH

Today's good workbench is inexpensive, simple, and assembles easily. It comes apart easily, too. Mack truck construction belongs in a Mack truck; you should think in terms of living room or, better yet, house-trailer construction—a minimum of materials, used well. In modern electronics a complete computer can get lost in your pocket. The man who's up-to-date thinks small and saves time and money at it.

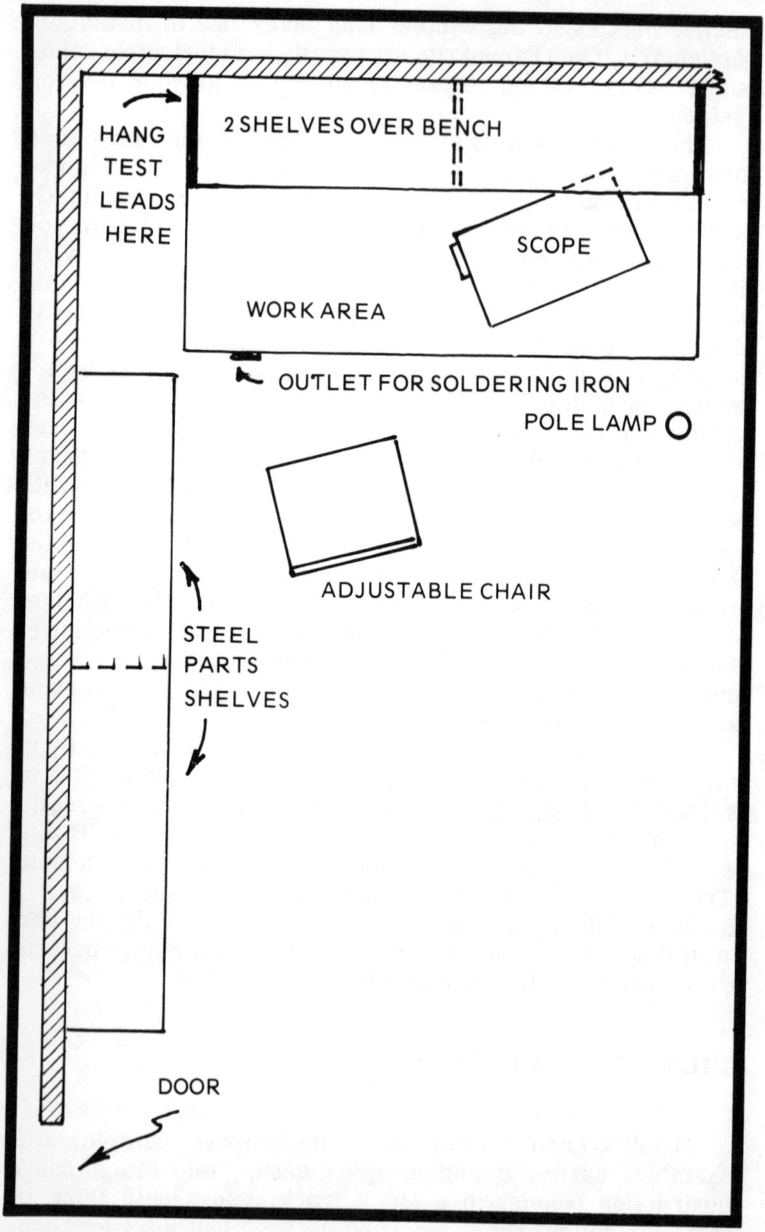

Fig. 2-1. A simple drawing at about one inch to the foot helps visualize your lab before you spend money on it. Try several before doing any work.

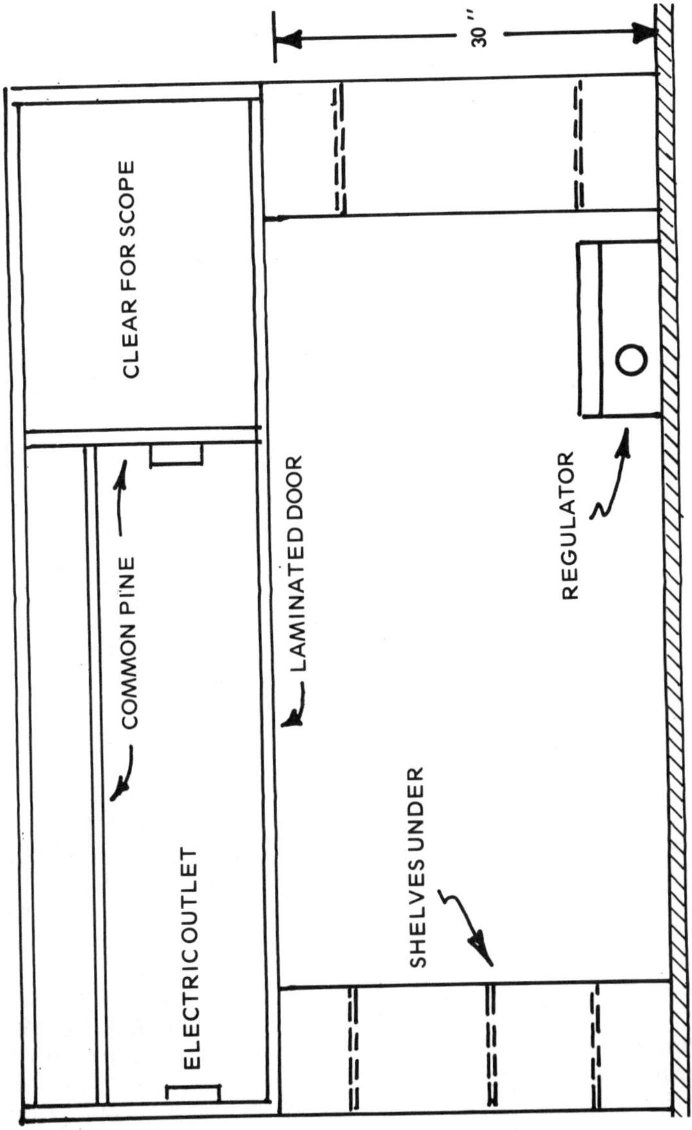

Fig. 2-2. Suggested workbench layout for a southpaw. If you're right-handed, make a mirror image or something like this. A complicated design probably won't be satisfactory. Set the left-hand bottom support in 6 inches and the right-hand support in 18 inches. This stiffens the bench and provides areas under the ends for storing boxes.

Two laminated doors will assemble into a very adequate, satisfactory workbench. They're lightweight, inexpensive, and they are strong enough. See Fig. 2-2. You need a door near 30 inches wide for a bench and another one, the same length, near 12 inches wide to go over it as a shelf. If you want to sit at this bench, your working surface should be near 29 inches off the floor if you're short, or a bit over that if you're tall. For sitting you need a secretary's adjustable chair and you can find a good enough used one for maybe $10. It will last years, although it may seem slightly disreputable. For standing height, think in terms of 36 inches off the floor; then you can use a draftsman's stool or one of the kitchen chairs available from Sears Roebuck. If you can, buy used; buy carefully, since these chairs are not adjustable like the secretary's chairs.

Shellac makes an adequate finish, and you can put a dark stain under it. Buy good supplies, and follow instructions. Two coats of plain floor enamel applied over shellac primer makes a nice job. Polyurethane varnish makes a good finish, too; be sure to use with lots of ventilation. (Considering its odor, you won't need much of a reminder!) Apply one coat, smooth with fine sandpaper and put on a finish coat. You may want to use a piece of linoleum, unfinished hard masonite, or formica, maybe 20 x 30 inches, to keep solder off the bench. My own style is to avoid splattering by soldering well. And a handy piece of scrap wood serves as a backing for mechanical operations to avoid scarring the work surface.

Setting up your bench is a problem in good design. Be inventive. The most satisfactory scheme I have tried is a couple of little bookcase-like assemblies as shown in Fig. 2-2. The bench simply rests on these without mechanical attachment. In a house trailer or aboard a boat, you'll need something more cohesive. An interior brace goes about 40 inches from one end, dividing the bench surface into working and nonworking areas. The setup shown in Figs. 2-1 and 2-2 is designed for a left-handed, worker but changes for right-handed users are easy to make. Notice this scheme provides for a very large scope, which you are likely to have if you are a serious worker. If you buy carefully and all new stuff, I think you can get a good bench into your lab for about $20. As long as you work in modern electronics, it will be enough.

LIGHTING AND POWER ARE IMPORTANT

Your lab needs electric power. One wall outlet is enough to manage, but it won't work out well. So we'll assume a more ambitious course. If the following seems elaborate, remember you can make up these circuits from lamp cord and five-and-dime surface outlets. In fact, that's just the way I did it in my present lab setup.

Your lab's electric power goes three ways: lights, bench power, and utility outlets. Good planning is to keep these separate. If a fuse blows it probably isn't your lights, and at a time like that you'd like to have those lights. If something plugged into your bench outlets fails, you want your instruments to stay on, since it may take some of them a little while to stabilize again after being turned off. So if you can manage it, you want separate lighting, instrument power, and utility outlet circuits, with appropriate fusing. Where all electric power comes from one source, such as a cable tapped into a box, branch connections and fuses can be placed in a utility or distribution box situated conveniently, like another instrument, somewhere in the room. Now let's look more closely at each circuit branch.

Lighting

Lighting is crucial. Men studying men working under various lighting conditions found what you probably know already. Poor lighting slows work and tires you faster. See Fig. 2-3. Good lighting is easy to manage today. You should use fluorescents if you can find any, since they provide a better

Fig. 2-3. Bad lighting can produce glare, insufficient illumination or both. Three hundred watts of incandescents will cost about 1 cent per hour and is worth it. Fluorescents are better.

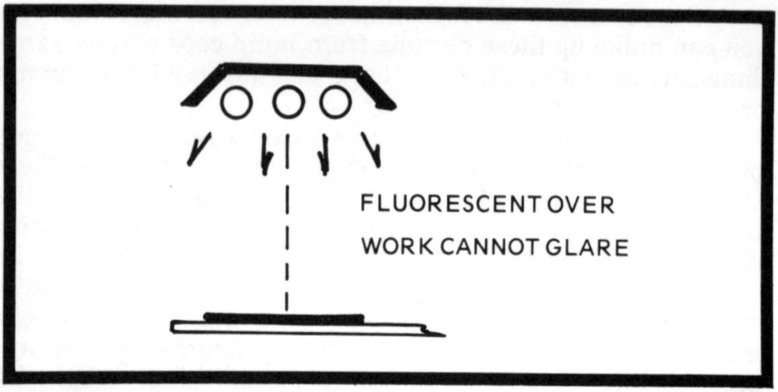

Fig. 2-4. This simple installation of two or three 40-watt fluorescents provides the very best lighting effect. If fixtures won't fit, investigate fluorescent circuit design and build your own.

light and use one-third as much power as incandescents for the same amount of illumination. Electrical noise is said to be a problem with some fluorescents, but I have used them under many different conditions, including some very demanding ones, and I haven't come across this. I recommend fluorescents over anything else.

A minimum setup is two or three 40-watt fluorescent lamps over the working area, about even with the bench edge or slightly back from it. See Fig. 2-4. If you search around a bit in some shops and industrial electrician's businesses, probably you can find a dirty but very adequate fixture for around $5. Usually this is one removed from a plant in a shotgun replacement program. They are often picked up by people who make repairs as needed and then sell them. Avoid anything that seems to have water damage or has been exposed to outdoor storage; it'll be rusty. Plug it in to see it work, and evaluate the noise it makes. If it works and is quiet, it should go for years. Current drain on your AC power line will run near 1 ampere.

An alternative choice, workable but far less satisfactory than a good fluorescent fixture, is a pole lamp right by the bench. It goes at the right if you're left-handed, and should be very near the bench. This pole lamp needs three 100-watt or larger bulbs in it and you may have to look around a bit to find one that will stand the heat these bulbs release. The key is air circulation. The best lamp setting usually is realized with the top-most lamp pointed down at the work and the other two aimed at the ceiling for general illumination (the ceiling should be light-colored). Notice that a pole lamp will use about 300-450 watts compared to the fluorescent's 120 maximum, yet it won't do as good a job for your eyes.

Instrument Power

You'll need an AC circuit along your bench to power your instruments if they don't run off batteries. Likely gear on this circuit is an audio signal generator, a capacitor checker, a soldering iron (**Don't** use a gun!), a VTVM if you have one, and the best scope you can find. The supply fuse is adequate for this line since most gear comes with its own fuse; if you discover you have something without a fuse, stop all work until this very real hazard is corrected. If your scope is a big vacuum-tube item, your instrument power line is heavy-duty to the scope outlet and lightweight beyond that. See Fig. 2-5. A Tektronix lab scope may take about 800 watts, and such a scope is no dream item for your lab. This kind of gear is available in the surplus market and is a good purchase when you buy knowingly.

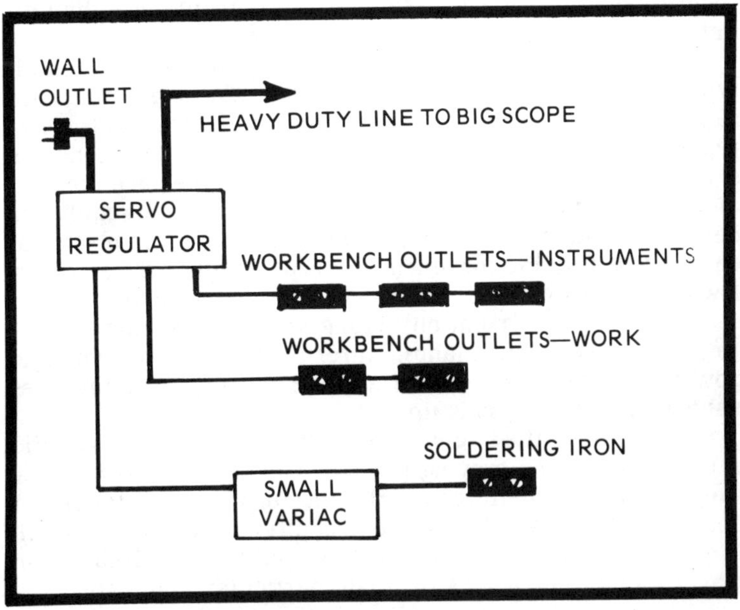

Fig. 2-5. A practical minimum for wiring around an electronics workbench. Use heavy wiring where needed and conventional lamp cord and surface outlets elsewhere. The voltage regulator is optional but desirable.

Utility Outlets

Work on the bench is power from utility outlets, unless it is run off batteries. For industrial work, and maybe for some radio amateur application, the utility circuit should be wired with heavy duty cable. For modern electronics work using small transistor circuits, this line will hardly ever be asked to deliver so much as an ampere.

Voltage Regulation

Today's power line voltages tend to be "drifty." A comparison of the estimates of present and future electric power requirements to the power companies ability to supply this power is not at all promising. And things probably will get worse. Therefore, a good voltage regulator in the instrument and utility supply circuits will prove at least a very nice convenience, and it may be worth more than that. When your lab has been working a while and has more or less amortized itself to a point where you are thinking about conveniences (careful; they consume money and divert attention from real progress), a voltage regulator has more going for it than an impedance bridge or a sweep generator.

There are three basic types of voltage regulators, and all types are available in the surplus market. These are magnetic regulators, low-harmonic magnetic regulators, and servo regulators. Of these the low-harmonic magnetic regulators are simple but the heaviest, and the servo types are the best all-around regulators. The less expensive magnetic regulators deliver an output distorted enough to trouble some electrical equipment. Buy the low-harmonic type. Regulators at really low prices may be 50-Hz units, which are no good at 60 Hz, or they may require or put out wrong voltages. When buying a regulator, check the manufacturer's tag for rated voltage, power and frequency. If the appropriate tag or assurances cannot be found, pass it up.

Some regulators generate a lot of noise. Magnetic regulators have no moving parts or electronics but the noise they make is sometimes audible at surprising distances. A servo regulator, which has a control circuit that feeds a small motor to crank a Variac up or down, may be rewired to improve its characteristics for your particular application.

PARTS MANAGEMENT

The common electronic parts are resistors, capacitors, inductors and transformers, diodes, triodes, special devices,

ICs, construction hardware, and assembly hardware. Any of these categories has at least several members. Each electronic component class has thousands, not including individual manufacturer's variations, and device differences. Managing these parts sensibly is particularly tough for the amateur who often has only one or two of a particular item. And to save money you may have to buy a batch of things, not all useful, such as a pile of circuit boards or a bag of miscellaneous semiconductor devices.

If you mean to avoid losing time and money obtaining small parts from a commercial supplier, or collecting such a pile that you cannot use your stock effectively, then you must come up with a really good approach to this parts problem. How do you keep things in order so you can find what you want or know that you don't have it? Reason wins: It can be done well and fairly easy, but you must be tough-minded concerning things for which you have no immediate use, yet not too inflexible in managing the now considerably simplified problem.

We begin with a general description of the well proven nonsystem described here. There are three parts to this scheme. First, parts and materials not evidently useful go into a large, out-of-the-way box, to be scrounged, re-evaluated, cannibalized, and eventually discarded at odd moments. A new batch of circuit boards, for instance, would go into this box. From time to time you may strip out the useful items, and then throw away the remainder.

Second, have some system of arranging things sensibly in not too great an area. Whatever the system, it should feature really fast and easy accessibility and some kind of order, so that when you want something you know where to look, and if it isn't there you know you don't have it. There are lots of ways to achieve this. To save space a given container may hold several kinds of parts, but all should be alike somehow. Other people may find your labels cryptic but you know what is to be found under each.

And finally, be tough-minded. There was a time when anything that had once seen the hand of man was rightly stored away for possible future reference. Today it is entirely too easy to let yourself slip to a point where some realistic observer wonders if you keep the property or the property keeps you, there is so much of it. Today you can usually get new things quickly, so you can be a little sticky about what you do or don't keep. Items that haven't a quite definite utility go out. Over the long run this does wonders in keeping your lab lean and active.

Magic Numbers

The key to effective component organization is the almost universal system of electronics "magic numbers." It is explained in the Appendix, and you ought to understand it well. This system is composed of two sets of numbers, the 10 percent and the 5 percent series. Memorize at least the 10 percent set. These are: 1, 1.2, 1.5, 1.8, 2.2, 2.7, 3.3, 3.9, 4.7, 5.6, 6.8, 8.2. Read this a few times to catch its interesting rhythm. If you like numbers probably you'll memorize the 5 percent series too, but this gets little use in amateur electronics.

Next, you must understand color coding. Most component values are color coded about the same way, and the individual colors always mean the same. If you can recall the color sequence black, brown, red, orange, yellow, green, blue, violet (sometimes called purple), gray, white; silver 10 percent, gold 5 percent, you have it. These represent the digits 0 through 9, with the two most common tolerance values. When no tolerance is coded on the component, it's usually 20 percent. In your purchases, aim for 10 percent or better.

Bin Storage

The best storage scheme I have found is a set of boxes divided into bins. There are a lot of these on the market, but few are good. Far better than anything else are the Vlchek P812 box, 9 inches x 13⅛ inches with 12 bins (Fig. 2-6) and the Vlchek P824 box, same overall size, but 24 bins. At this writing, these boxes are head-and-shoulders above anything else for overall value and utility, and they are moderately priced. Unless you have an unusual supply of space and money, each Vlchek bin should do double duty or better. Little slips of paper in some bins concerning the components are very helpful, and this keeps specs with the appropriate parts. For the most used electronic components (resistors, capacitors, and inductors), you can get triple use from your bins.

Bin System

I rate resistors in ohms, capacitors in picofarads, and inducators in microhenries. At the low end a 3.3-ohm resistor, a 3.3-microhenry inductor, and a 3.3-picofarad capacitor would all turn up in the same bin. A 1-millihenry choke is 1,000 microhenries, and turns up in company with the 1,000-ohm resistors and 1-nanofarad capacitors. A small 1.2-henry inductor keeps company with some 1.2-megohm resistors and maybe a 1.2-microfarad capacitor. If you're a beginner, this system will help you develop a much needed feeling for component parts values.

But not all small parts fit the 10 percent system perfectly. How about all the 5 percent parts? The rule is a bin gets any component of its labeled nominal value but **below** the next bin's value. The 3.3 bin also gets the 5 percent 3.6 value or any oddball under the next standard or, in this case, 3.9 value. With all parts of one kind in the same place, the bin system simplifies purchasing policies. When a bin gets too full or empty, needed changes are instantly apparent.

Semiconductors

Semiconductors include diodes, triodes, special devices, and ICs. If you think diodes and triodes are big things with vacuum inside, that's not so. A diode is an electronic device with two leads; a triode has three leads. An ordinary transistor is a triode and goes in the same box as an SCR, which is a kind of Shockley diode. But the Shockley diode goes in the diode box with the other diodes even though it's really a PNPN device more complex inside than the triode transistor. "Special devices" is a catchall label that goes on its own box,

Fig. 2-6. A 12-bin parts box can do double or triple duty by storing resistors and capacitors, and maybe inductors.

and digital and linear ICs get one or two boxes, depending on the variety you have on hand. Vlchek P824 boxes serve for all semiconductors. Special devices such as unijunction transistors, diode capacitors, thermistors, and other special units may go in either the special devices box, or be classified by the number of leads and sorted into diodes, triodes, and ICs. Notes kept in the bins include suppliers addresses for future replacement needs.

Other Bins

Vlchek boxes are best, but there are other resources. A bin is any useful container you can store parts in. It shouldn't be too much larger than the contents, or you waste a lot of space. Soup cans, painted and labelled, make fair bins but it is hard to see inside. Country cats thrive on dog food, plus what they find outside; maybe you can find bins there. Paper companies supply a variety of inexpensive containers that will be long-lived, since each gets relatively little use. Cigar boxes work for small parts collections, but are hard to find.

So-called "utility bins," available from very many suppliers, are usually very light steel boxes with many plastic drawers in them. They are not very practical. Before buying, you should see the bin and then get one to try it out. It may not look as good in your lab as on the showroom floor or catalog page. Whatever you choose, don't buy a "general-purpose" thing with assorted size bins in one box or drawer; these simply don't work in any sensibly organized system. And because all electronics classification systems are based on 12 values per decade, boxes with five or ten bins also are not workable.

Assembly Hardware

This category includes a wide variety of small parts including connectors, knobs, nuts and bolts, lockwashers, solder lugs, lug strips, flea clips, internally threaded spacers, pilot lamp assemblies and bulbs, and other small items. Practically all your assembly hardware can be stored in a Vlchek P824 box or two, and if you're careful you can do without a P812 box for larger pieces.

There is a fantastic variety of nuts and bolts available for assembly work. These are easy to find today, so a very large stock even of the most used sizes isn't necessary. The most useful bolt style is the fillister-head, which is as good mechanically as the round-head but looks better on surfaces and panels. The sizes listed in Fig. 2-7 are the most used, but good planning can hold the emphasis narrowly to the 4-40 and 6-32 sizes, with a few 8-32 types on hand.

Catalogs

The very best stock parts collection is the one that ties up the least money but most often has parts when you need them. This means you carry a practical minimum of most used parts, know who the best local suppliers are, and keep a really good catalog collection. Experience soon shows who the best mail-order people are, and which needs each can most successfully meet.

Major suppliers today are Lafayette, Burstein-Applebee, Allied Radio Shack, Cramer Electronics, and a number of regional operations. You can zero in quickly on these by checking the yellow pages for your phone district and for a nearby large city. Many of these companies will provide you with catalogs; some are industrial outfits not interested in doing business with individuals. Not only do catalogs tell you what parts you can get, they often tell you (more or less) what the electrical properties of these parts are. This is extremely valuable reference material. Older Allied Electronics catalogs, and to a lesser degree Allied Radio catalogs, were

	1/4"	3/8"	1/2"	3/4"	NUTS	LOCK WSHRS	SOLDER LUGS
2-56	FEW				FEW	FEW	
3-40	FEW	FEW			FEW	FEW	
4-40	FEW	FEW			MANY	MANY	
6-32	MANY MANY	MANY MANY	FEW	FEW	MANY MANY	MANY MANY	MANY
8-32		FEW	FEW		FEW	FEW	FEW

Fig. 2-7. Use this chart to estimate how much nuts-and-bolts hardware you really need. But fillister head types only in nickel-plated steel.

Soldering iron & variac; solder	Electrician's knife
Screwdrivers	Slipjoint pliers
Needlenose pliers	Tinsnips
Diagonal cutters	Vise
Insulation strippers	Electrician's tape
Hand drill & drill bits	Hacksaw
Punch, scriber & ruler	Small files
Wiring aid	Aluminum-oxide paper
Reamer	

Fig. 2-8. Here is a list of tools you'll need. But buy very carefully and choose the best quality you can afford. Most of these have a useful life of several years.

exceptionally good in this respect. At this writing the company has just changed hands and this thorough-going workmanlike policy of supplying facts may not continue as the company's style.

GOOD TOOLS

Three things in amateur or professional electronics are somehow alike. One is ideas—tools of the mind. Another is instruments—tools for getting facts concerning events in electronic circuits. And the third, discussed here, is just plain tools.

It takes no skill to buy tools, simply money. Be advised that half or two thirds of the tools offered to electronics amateurs are without redeeming value, regardless of what the advertising says. To use your tool dollars best, buy a basic minimum outfit to start with, and then buy more when it hurts to do without. Following this policy will keep you away from an expensive booby trap too clearly set for beginners. The basic set listed in Fig. 2-8 will get you through the lab work and projects described in this book. You will have to be tolerant sometimes. But it's enough for building a computer or other elaborate electronic system, if that is how your interests run.

Soldering Iron

Never buy a soldering gun; if you have one dispose of it. Each time you use it you must wait for it to warm up, and a short time later it is too hot. You cannot work precisely with it. The recommended outfit listed in Fig. 2-9 has proven adequate over many years experience. It seems to be a luxury, but an SCR controller or a variac is a practical necessity for the best work. The variac is preferable, and will do many other jobs around the lab.

Solder

Use only rosin core solder. The acid from acid core solder continues working until the joint is destroyed. All manufacturers' warranties are off for any gear subjected to this poisonous stuff. Buy light gauge solder made by a good manufacturer. .060 diameter solder is pretty big these days; .020 to .030 inches diameter is good and some workers choose smaller than this. The solder's tin-lead composition is marked on the package, tin first. Look for 60-40 solder, which is 60 percent tin and 40 percent lead. This costs more than 50-50 but has a lower melting point, which is good for today's tiny components. You may come across 63-37 eutectic (lowest melting-point possible) solder, but this is not markedly better than 60-40 solder. If you want to work in an instrument, such as a Tektronix scope, which seems to have solder directly on ceramic, be sure to use the special silver-bearing solder specified by the manufacturer. Ordinary solder may weaken and break the solder-to-ceramic joint. Look out for a retailer who sells 30-70 solder, apparently banking on your not knowing the difference. Of course it's cheap—not enough tin in it. And be sure the solder has a **rosin flux core.**

Screwdrivers

You need at least two screwdrivers: one that has a blade near one-eighth inch wide, the other with a quarter-inch blade.

Ungar 776 handle
Ungar 1233 chisel tip
Ungar 1237 thread-in element
Ungar PL-111 pencil tip (bend to 30-degree angle)
Light-duty variac of any type

For most work, a below normal variac setting is best. If some heavy soldering needs to be done the variac is turned up to 130 volts using the 1233 chisel tip.

Fig. 2-9. Recommended soldering iron outfit. It is appropriate for nearly any amateur electronics work. Stay away from soldering "guns!"

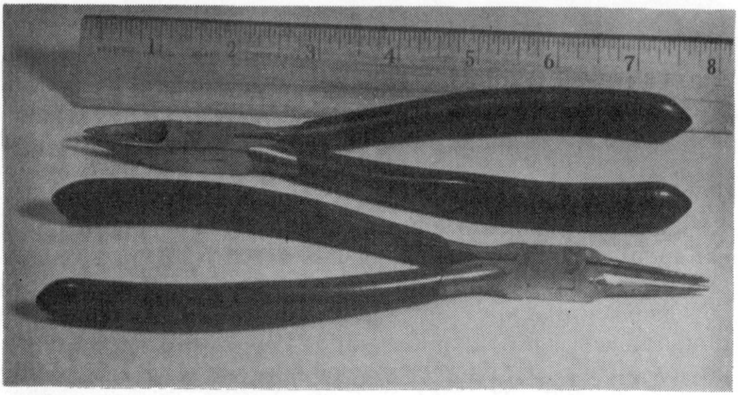

Fig. 2-10. Good pliers and diagonal cutters have long handles and are not too small overall.

Learn how to sharpen screwdrivers right; this means flat sides and sharp, square corners. Screwdriver handles should be fairly large and ribbed; the shafts near six inches long and made of good metal. Sears Roebuck's good grade screwdrivers have worked well in my experience; some companies sell sets of very crummy screwdrivers at apparently economical prices.

Needlenose Pliers

Good needlenose pliers come down to a rather fine point (Fig. 2-10) at the working end, but have handles at least 8 inches long. These are fairly hard to find. Suppliers seem to think pliers for fine work must be small on both sides of the hinge. This is false. Your hands are designed for relatively large movements but have excellent sensitivity to small forces. Good pliers are large at the hand end but small at the work end; if the hinge is free working you can very well feel what is going on and the needlenose pliers reduce your natural hand motions to what the work needs.

Diagonal Cutters

The same need for a long handle, free hinge and short work end applies to diagonal cutters (Fig. 2-10). They can be used for moderately heavy work, if whatever is to be cut is carefully placed near the hinge. Shop around; seek tools that work well and are obviously made of good materials. If they look like the manufacturer values them, maybe you will, too.

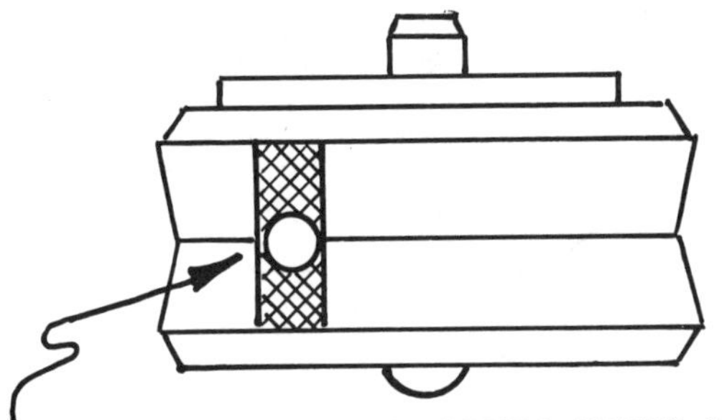

Fig. 2-11. Some careful work with a fine file converts an inexpensive fingernail clipper into a superior insulation stripper.

Insulation Strippers

A pair of fingernail clippers can be modified into an excellent stripper by careful notching with a three-corner file. See Fig. 2-11. If your first try fails, enlarge it for bigger wire and try again on the other side. Another choice is the adjustable stripper made by Miller. You want one without a spring, or you can take a spring off if it has one. It interferes with your feel, which is needed to avoid cutting into the wire. Careful grinding pares off unnecessary parts of the stripper, making it much nicer to use around circuits. Before buying, check to see that the notches meet properly for a square aperture, and then adjust for stripping no. 22 wire.

No. 60. Utility small; wires through PC boards
No. 52. Small flea clips like Vector T28
No. 42. Large flea clips like Vector T9.4
No. 34. 6-32 tap drill
No. 27. 8-32 tap drill
No. 25. 6-32 clearance drill
No. 16. 8-32 clearance drill
¼". Utility fraction drill; larger sizes optional

Screw tap and clearance sizes are chosen large for convenience in hand-tool work.

Fig. 2-12. Table of recommended drill bit sizes. This set is adequate for practically any work.

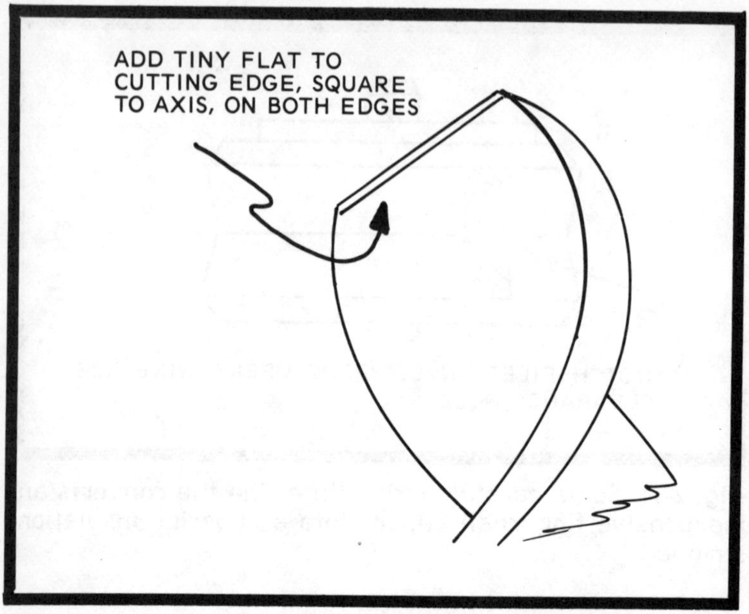

Fig. 2-13. A drill bit can be modified to prevent hogging into brass or plastic. This must be done precisely using a very fine carborundum stone.

Hand Drill & Drill Bits

Many modern electronics construction materials come full of well chosen holes and few new ones are needed. Yet sometimes there is drilling to do. A hand drill does this nicely but an electric drill is somewhat more than a luxury. It does not need to take over one-quarter inch drill bits. Drill bits come in two size systems: fractional inches, and machinists' wire drills. Only one-quarter inch, and perhaps five-sixteenths and three-eighths inch fraction bits are needed; all the rest should be machinists' bits. See the table in Fig. 2-12. Expensive bits aren't needed, but you should be certain you buy good high-speed steel. Try Sears Roebuck for these. A set is nice, but won't really pay for itself. A few passes with a very fine carborundum stone will stop these bits hogging into the work, which can be startling if it does no more harm than that. See Fig. 2-13. This does not really dull the bit, and greatly improves its performance in certain plastics and brass-like metals. Check a machinist's book, or have an expert show you how to modify the bits.

Punch, Scriber, & Ruler

These tools are used for marking material that is to be bent, cut, or drilled. Very often you work by eye or use materials made to your order; in modern electronics you do little mechanical construction. The ruler needs a steel edge; an office model will do nicely. The punch can be shaped from a nail, if you use it gently on soft materials like plastic or aluminum. If the punch is very sharp, it can do double duty as a scriber.

Wiring Tool

A variety of wiring tools is available, featuring assorted ideas about something for holding wire at one end and something for prodding into solder joints at the other. These ends should be small and sharp, sized appropriately for work on small joints using no. 22 solid wire. Some wiring tools on the market are far too big.

Reamer

A reamer is a simple, useful or necessary tool that makes little holes into big ones. It equals or surpasses the five-sixteenths and three-eighths inch drill bits for many jobs since the finished hole size can be adjusted exactly to need. For instance, you can ream out a hole for friction mounting an odd-sized IC or transistor socket. The reamer should be used gently. If used too forcefully, it cuts out a hole with many little flat sides (try this) and the hole size cannot be accurately controlled. Cutting slower takes very little more time and gives you nice circular holes. You may want to clean up the edges of the hole after you have done the cutting.

Electrician's Knife

This is simply a jacknife with a locking screwdriver blade you'll hardly ever use and a very good steel blade you will use often. If this blade is not abused, it will keep an edge through many lab jobs, improving metal and plastic pieces.

Slipjoint Pliers

You need pliers to hold things. To spoil needlenose pliers quick, try twisting with them. The less precise slipjoints are far tougher. Choose small ones, near 6 inches long and with jaw width of about three-sixteenths inch.

Of course you'll want more tools than these. To choose, read catalogs and visit stores, cultivate a general familiarity with many types of tools, and talk with people who use tools. When you become really convinced you're hurting without something and that you have finally chosen the best purchase available, put the matter aside a couple of days. If you still think you need it after that, then buy it. This deliberate, informed approach will cut your tool budget by half or better, but you will still spend plenty on tools.

TEST GEAR: A KEY INVESTMENT

When buying test gear for your lab, be choosy, and don't choose very much. Many instruments and combinations of instruments are on the market; your best bet is the simplest, most appropriate of these. Later, when you think you want more, think again. Maybe you can improve your technique or your idea background, or design some adapter circuit so that you get by without a major new purchase. The following list is arranged in order of joint utility and affordability (reciprocal of cost).

Multimeter

A multimeter combines in one instrument the circuits of a DC voltmeter, an AC voltmeter, an ohmmeter, a milliammeter, and perhaps an ammeter. There may be other circuits too. It competes with the oscilloscope in general utility but comes first in this list because it is far cheaper. A good multimeter should be your first instrument choice.

For maximum utility it must have a fairly large scale (a mirror in low-priced gear is a sales gimmick that wastes needed scale space), a low-voltage range not exceeding 2 or 3 volts full scale, and a sensitivity of 20,000 ohms or more per volt. That is, the meter's maximum drain from the circuit being checked should not exceed 50 microamps. One suggested multimeter is the 30,000 ohms per volt model priced under $20 from Lafayette, or a smaller one priced just over $10. The Lafayette catalog assigns no model or type designation to these meters. Such inexpensive multimeters are not very accurate, but they are adequate for good work. More expensive models are available up to $100 or so, but should be reserved for professional work. A multimeter will be long-lived if not abused. Its only really electrical part is a battery or two, which should be replaced whenever the ohmmeter cannot be zeroed. If the instrument is stored the battery should be removed.

Audio Oscillator

An audio oscillator is a very versatile instrument, valuable for many applications beyond checking audio-frequency amplifiers. A square-wave feature has little value, since the knowledgeable worker observes frequency response more usefully by testing at low and high frequencies. The frequency range should extend to at least 100 or 200 kHz, since these higher frequencies are needed for checking the performance of feedback loops. A nice utility oscillator from General Radio that tunes up to 1 MHz may turn up in a surplus store; if so it can save purchasing an RF signal generator.

The audio oscillator replaces the RF generator if you add a circuit like the thin-line pulse generator described in Chapter 10. For pulse work a simple adapter circuit consisting mainly of two decade countdown ICs converts a 20-Hz to 200-Hz sine-wave input to a stable one per five seconds to 2000 per second pulsed output. This is about right for most basic logic circuit work.

At this writing only one audio oscillator on the amateur electronics market deserves attention. This is Lafayette's general purpose sine-square-wave generator priced near $40. An alternative is constructing the audio oscillator described in Chapter 10. This instrument's fix-tuned operation is far more successful than you might expect.

Oscilloscope

For amateur electronics use, an oscilloscope should feature direct voltage reading capabilities and have a triggered sweep. A delay line is nice for pulse work, but a convenience only, since a pretrigger to the scope will do the same job. Inexpensive service-type scopes picture repetitive waveforms by using a recurrent sweep. That's not enough for modern electronics work where pulse-type signals appear on an irregular time schedule. Radio Electronics and Electronics World magazines have published descriptions of triggered-sweep adapter circuits that can be used with service type scopes. If you are working under a really minimum budget, look into these.

DC and AC voltage level information is very valuable, but can be read only from a calibrated screen displaying the signal at a known sensitivity in volts per division. "Calibrators" are available or can be built to display a known signal on the scope screen before observing another signal of unknown size, but this is far less convenient than working with a calibrated scope.

An appropriate scope can be purchased new, used, or surplus, but if you are located near a large city, give a thought to renting a scope. See phone book yellow pages to find an appropriate company. If this utility is available, make the most of it to zero in on what other gear you really need. New Tektronix scopes are very high priced but worth it. Used Tektronix scopes are available in the surplus market at $100 up. Since the Tektronix line changes slowly, recent catalogs are a guide to older gear. Tektronix follows a very tough-minded policy about manufacturing the best gear the state of the art permits, and all their machines were the best anywhere when built. A very dirty Tektronix can be cleaned remarkably well. Contact a local service office for their comments on how to do this; it could be worth your while to ask them to give your used Tektronix the complete treatment. Your scope deserves this heavy-instrument policy because its tremendous flexibility of application is literally beyond description, and because money invested in a scope dissipates slowly. Eleven hundred dollars spent on a car (except a Volkswagen) is soon gone. The same amount invested in a good scope will outlast even the VW by 3-to-1 or better.

VTVM

Many VTVMs (vacuum tube voltmeter) have no vacuum tubes in them. They use field-effect transistors. The electronics in the instrument amplify a tiny drain of a few tenths microamp taken from the circuit being tested and display the result on a large meter. The VTVM probe contains an isolating resistor, so the tested circuit experiences inappreciable signal as well as DC loading. Its ability to look into a working circuit, even a very critical one, without upsetting anything makes the VTVM a valuable test instrument. All the low-priced VTVMs on the market are about the same. None are highly accurate, but great accuracy isn't needed for most VTVM test applications. Some industrial VTVMs now available in the surplus market are extremely nice instruments. Older, very large ones with small meter faces, should be avoided.

Capacitor Checker

Now we come to the less needed but still highly useful instruments. Most of today's capacitors are labelled as to type, capacitance, and other details, but it is still useful sometimes to actually test for capacitance, leakage, and for electrolytics, general health. A shiny new capacitor checker may be a bit hard to justify, if you get your electronic parts from reasonably good suppliers (it's worth the slightly greater cost). But if you come across a clean used one and you like the way it works, you will find it a convenient tool occasionally.

You can gain much practical experience and meet some valuable new ideas by making one of the many capacitor checkers and bridges described in Radio Electronics, earlier issues of Electronics World, in QST Magazine, and in that well run up-and-coming magazine, Ham Radio. A simple wheatstone bridge circuit will do nicely for most work around the amateur electronics lab. The Sprague "Telohmike" capacitor checker, presently priced near $150, is a bad buy at any price, new or used. It is difficult to use and has no practical features not available at much lower prices. A clean Lafayette TE-46 capacitor checker, if you can find one, should prove a very good instrument. It may need a new leakage meter. Apparently Lafayette no longer sells this instrument.

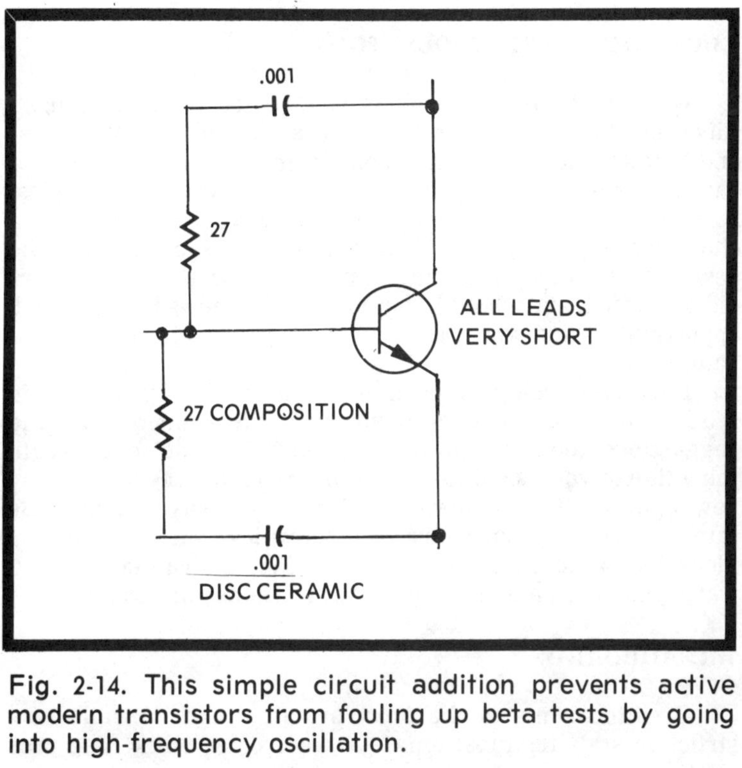

Fig. 2-14. This simple circuit addition prevents active modern transistors from fouling up beta tests by going into high-frequency oscillation.

Transistor Tester

Heathkit's transistor tester is probably the nicest available, if you must have a transistor tester. This instrument reads DC beta directly, but if you want to know DC beta it is quite easy to design a circuit that will feed a few fixed currents in the microampere range into the transistor's base and indicate the resulting collector current. Although it falls at the bottom of this list of suggested gear, the Heathkit transistor tester is definitely a nice item to have around sometimes. If a transistor tester gives erratic results, this may be due to the fact that the transistor is going into oscillation at very high frequencies. The transistor oscillates because it is biased into its active region during the test, and some modern transistors are very active indeed. The transistor sees the tester's wiring as a resonant circuit, but this can be stopped easily. The circuit of Fig. 2-14 is installed right at the transistor socket, with very short leads; it serves as a high-frequency load, damping out oscillation. This does not affect normal tester operation in any way.

GOOD IDEAS ARE TOOLS, TOO

Whoever thinks of **ideas** as electronics lab furnishings? Any professional engineer does; that's why he's a professional and that's what makes him an engineer. Try to find ideas you can use again and again—basic thoughts and principles. Exercise your mind by seeing electronics as a game and a challenge to your ingenuity. See if you can second guess the news events you read about, working out what the engineers did or missed, and what all that means. The news is full of such opportunities and who knows, perhaps you could find one for yourself there!

Further develop your mental storehouse by just plain reading. Refer to the notes about this in the Appendix, and the suggestions about how to use a library. I cannot exaggerate the value of your local library. Library resources offer many new opportunities to discover things, yet save money. Be familiar with the hardware of electronics, but seek out the ideas too. Without good ideas your lab, instruments, hardware, your work and invested time and money are wasted.

BREADBOARDS

Breadboarding is the most practical and sensible construction style for most amateur electronics work. The pros use it a lot too, but most people see only manufactured gear and often try to imitate it. They don't meet the interesting laboratory originals. Breadboarding is fast and simple, requires few tools, and it throws the emphasis where most needed—on the circuit. Breadboarding for a month or two will do the "frozen style" advocate much good, if anything can. And the beginner who should meet many circuits as he begins to understand how things work gets results in a minimum of time and at least cost.

The worker who learns breadboarding and regularly works this way soon becomes aware of strong differences between various construction styles. He observes that some people build few circuits and spend a lot of money at it, but those circuits look very good. They are full of beautiful assembly, square lines, new parts, extremely clean métal, perhaps nice paint jobs or expensive commercial cabinets, etc. This slow, expensive style is the "frozen method" of doing electronics construction. The chief advocate of "frozen style" work is the American Radio Relay League, which publishes a regular magazine (QST) and an annual handbook (Radio Amateur's Handbook). Both are full of much neater work than most people can afford in either time or money. The other side of the coin is evident, too. How circuits work is a weak point in these publications.

The following examples are supplied for your benefit but you shouldn't copy them. Try your own ideas; you will more or less have to since not all materials seen here are available everywhere. (But they were purchased from major suppliers.) We begin with a plain wood breadboard.

Style 1: Here a piece of one-half inch plywood is improved by two rows of nails and a bracket for mounting controls. Dimensions are uncritical, but my experience is that smaller boards are usually handier than large ones. See Fig. 2-15. Two parallel bus wires supported on the nails cross the board to serve as supply line, ground, and for parts mounting. The working circuit is hung on air between these, but a heavy part such as a transformer may be held down by a woodscrew.

Style 2: A similar breadboarding system is based on a utility box. Various kinds of these are available with assorted perforated covers and they are very nice for fast circuit assembly. In Fig. 2-16 the Style 1 system is copied over with the supply bus wires supported on several flea clips. The controls (if any are needed) are mounted in the box, avoiding any panel construction, and the board is notched for wires going down inside the box. Batteries could be installed in the box, but this turns out to be inconvenient in practice. Since the

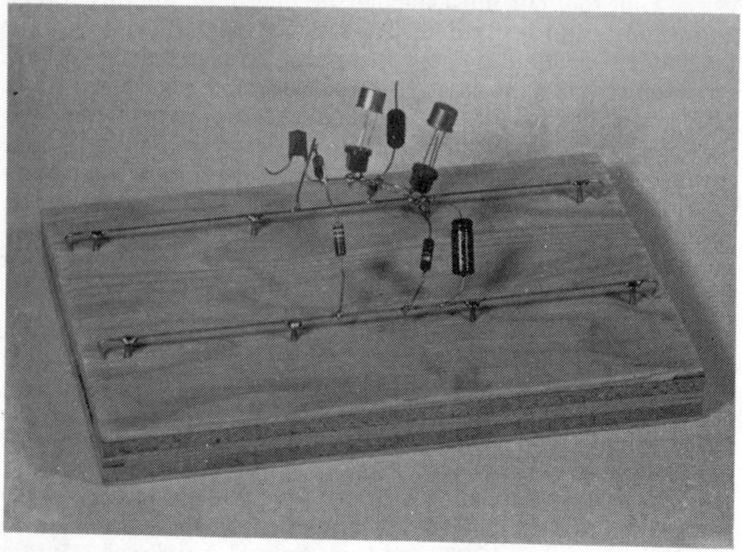

Fig. 2-15. Two rows of nails, 2.5 inches apart along a plywood board, makes an excellent modern electronics breadboard.

Fig. 2-16. A piece of perforated board and some clips makes a nice breadboard and offers a way to combine finished with trial circuits.

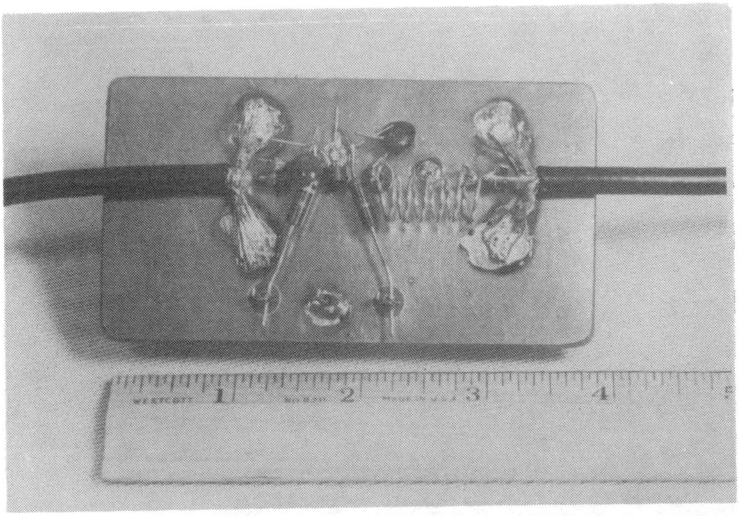

Fig. 2-17. This high-frequency grounded-base RF preamplifier was assembled in 15 minutes and produced good results on the 2-meter amateur radio band.

Fig. 2-18. The pulse generator discussed in Chapter 10, installed with its power supply, in a convenient aluminum base. Another circuit could be substituted for the pulse generator in a few minutes.

Fig. 2-19. The simple diode receiver is developed in Chapter 6. The small plastic box makes a very nice base for circuits that don't need shielding.

perforated boards are available without boxes, two or three of the boxes can serve for work with many circuits.

Style 3: The above ideas can be applied with variations to surplus perforated board, to printed circuits, to copper-clad boards made into high-frequency circuits and to many other phenolic board assemblies. The arrangement shown in Fig. 2-17 is adaptable to lab work or more permanent applications. Here, the base is a piece of aluminum shaped as shown in Fig. 2-18. A sheet metal shop can do these easily, but you should ask their price for the job before committing yourself to paying it. Some union shops are extremely expensive.

Several solder lug strips are mounted on the aluminum base, about a quarter inch outside the intended circuitboard location. A piece of cardboard, polyurethane foam, or plastic fits inside the lugs so the board cannot fall onto the aluminum base. Then the circuitboard or assembly is simply hung on wires between the lug strips, without any mechanical installation. This system is simple, easily takes new projects (key parts of older ones can be stored with notes in brown paper bags) and yet is inexpensive. Of the various styles described here, probably this is the most all-around useful one.

Style 4: Still another way to build simple circuits eliminates almost any need for tools. Allied Radio Shack offers thermoplastic "P-boxes." These boxes aren't as useful as the advertising implies, but they are still very good for quickly assembling simple circuits that are to be used a while. Usually, these boxes cannot be reused(see Fig. 2-19). The parts are mounted by sticking their leads through the holes in the box, bending the leads over, and soldering on the reverse side. The manufacturer's literature suggests twisting these leads together, but this is slow, clumsy, degrades the components, and makes changes difficult. The leads should simply be brought together by careful bending, and soldered. The example shown here is the evolved diode receiver described in Chapter 6.

Style 5: Everybody uses the "frozen style" sometimes. It is fine for circuits that are definitely going to be around a while. For instance, see Fig. 2-20, which is a vacuum-tube voltmeter made up from a rare but nice surplus meter for lab bench work. It is based upon 5 x 7 x 2-inch chassis, with a panel bent to a slight slant and braced with two strips of inch-wide aluminum. The panel has a one-inch lip at the top, which improves appearance and serves as a handle. All this construction was preceded by appropriate design and bread-

Fig. 2-20. This "frozen style" test meter is a good lab instrument. The wise amateur does little work like this because it is expensive and utilizes time poorly.

boarding, so the work went on a while. But at this writing it's done a good job for some five years, so the time and money invested is well amortized. The "frozen method" was justified, this time.

Breadboarding "lab systems" are available at prices amateurs can manage, but I haven't thought they were very successful. The Vector Corp., of Glendale, Calif., works on this idea and has come up with some practical systems. They also sell a variety of very sensible and useful construction supplies. With these examples before you, you can invent new styles according to need. These will depend upon what you can find or afford for construction materials. No matter how affluent you are, you should try a Style 1 project or two. Whatever styles you eventually settle on, they should feature maximum accessibility, convenience, simplicity, and minimum cost.

CHOOSING PROJECTS

A less formal title for this section might be "chunk theory." The field of electronics is practically without limit, but it's no good to jump in without some objectives or goals in mind. A year's output of periodicals, manuals, industrial publications and other resources contains a tremendous amount of idea stuff about designing and constructing circuits. Add to this library collections reaching back some 10 to 15 years and this is an example of the written resources available to the serious amateur. His own ideas will contribute still more to this, and it all somehow, has to be pared to one man's capacity. Chunk theory does this nicely. You simply bite off small sensibly sized chunks and avoid taking on new ones. When you have finished or otherwise disposed of work under way, then you look around, and you think a bit, and choose new work. This saves time since time is used better; it saves money because the money goes to some purpose rather than going out generally. Little experience is needed to size up a project, once it's generally defined, and to decide if it's good for now, or not now.

Let us look closely at a particular project. Consider the improved diode receiver from Chapter 6. A few details are immediately apparent, once the description is carefully reviewed. Certain tools are needed, but few. Some parts are needed, but only the diode and transistors might prove difficult; the diode ought to be mounted in clips. Is an adequate battery available? Come to think of it, are there broadcast stations near enough to be heard? (Probably yes.) What sort of antenna can be put up, since one is needed, and what about grounding? Would it be interesting to retune it for long-wave reception? Maybe it should be tried first, just as described for broadcast reception, and then retuned. (Yes, if you're new to electronics.) Sensible questions, sensible answers, and when you come into the middle part of the project, if you decided to do it, you're prepared to be effective.

The project's middle part is either do it, or drop it. Doing it begins with understanding the design or originating it, proceeds to parts and breadboarding, may go on to a "frozen style" construction job, and ends when what was to be achieved is either accomplished or (temporarily) dropped. Finally, the end part of the project consists of making a manual if the project is complex enough, of collecting notes and records, and writing in appropriate comments. This accumulated work ought to be adequate for building the circuit again, and it includes test results such as circuit voltages, frequency response, performance quirks, and so on.

All this sounds reasonably like "frozen method" work. It can be that, if it goes into complete detail. Experience will show that much detail can be omitted without weakening the key benefits of a systematic approach to the work: utility, reinforcement of new things learned, and maximum benefit from time and money investment.

Chapter 3

Useful Engineering Ideas

As any engineer knows, ideas are of central importance in electronics. Think about a circuit—any circuit. You cannot see it working. The flowing electricity, or the electricity available to flow, is invisible. Unlike machines, whose pistons and gears, levers, wheels, belts and systems make visible motions and generate sounds and other signs of life, events in electronic circuits are usually indetectable without special instruments.

Yet people can picture what goes on in electronic circuits. Anybody who masters the basic ideas of electric current flow (a very real kind of motion) and is familiar with the properties of electronic components can develop pictures in his own mind to replace the visible motions of machines. With practice, you can learn to read a schematic diagram and accurately picture what goes on in it. And you can learn to make up your own meaningful picture of what you want to happen, and then draw a schematic of the circuit that will do the job.

This chapter provides the basic ideas needed to understand electronic circuits—the basic ideas that replace the machinist's practiced hand on the solid parts he sees and works with. You will use these ideas, along with the results of simple tests and calculations, every time you think about electronic circuits. There is little true theory here—the ideas are all closely tied in to things you do with meters and batteries, and with some important closely related concepts about how circuits work.

There are lab experiments here, too. Some are so simple you hardly notice they're things to do rather than to read about. For maximum benefit from this chapter, study these little projects and carry them out, even if they don't seem to deserve the time. Each one you do gives you a clearer idea about some detail of how things work in real, complicated electronic systems. You will also find tests and experiments you do in your mind. Most of these are simple exercises resembling circuit design questions. For instance, you may begin with "What if...?" or "Suppose that..." and go on to

"Then...follows." Such thinking is actually a valuable lab test in effigy, and it helps you check and improve your feel for how and why things happen in real circuits. You might call them "thought experiments." You can do thought experiments anywhere, without any equipment beyond paper and pencil.

THE NATURE OF ELECTRICITY

What is electricity? In this book we sidestep many basic and extremely interesting questions about what electricity really is. That research leads into nuclear physics, cosmology, and the philosophy of science. It would make a very fine hobby in itself, if you like to read and wrestle with tough problems. But we shall stick to the hardware side, omitting basic theory almost completely.

Our definition of "electricity" is that it is basically an electric charge. This electric charge is something physical that may be stored in a capacitor, reside on an insulated object, or flow through a wire, a magnet, or a circuit. Precisely what an electric charge is, even on a practical level, doesn't concern us yet. We simply need some picture of how a charge behaves and what rules it seems to follow. Later we will go into more detail concerning what really happens inside circuits when a charge moves through them.

There is only one kind of electric charge. If there is an excess of it we say the charged object or terminal is

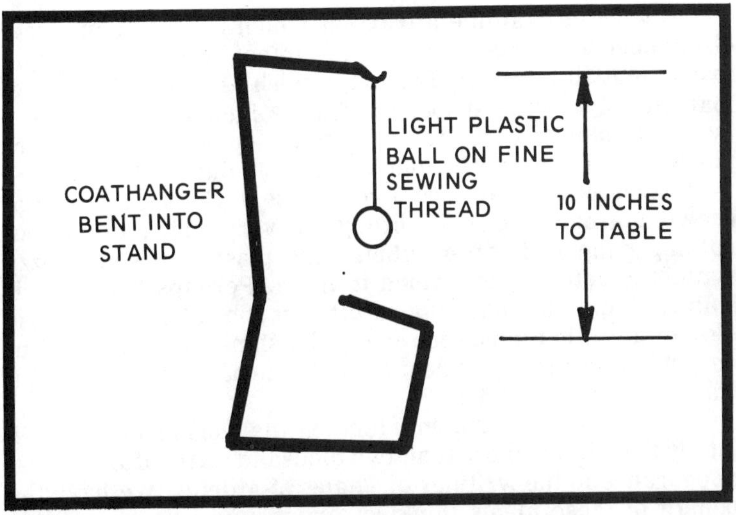

Fig. 3-1. A piece of plastic foam hanging by a fine thread demonstrates the effects of static electricity.

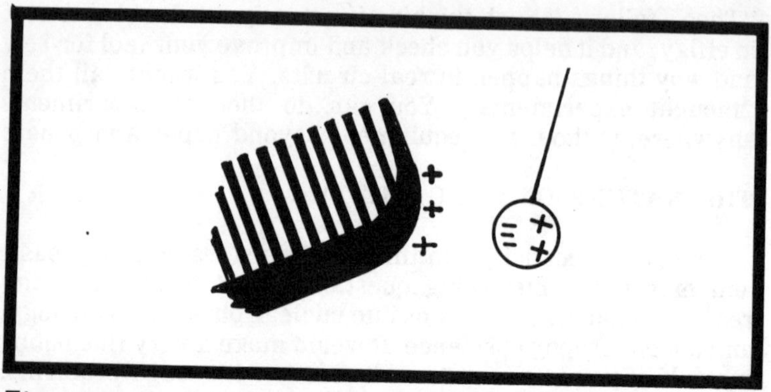

Fig. 3-2. A positively charged comb induces bound charges in the ball. The negative bound charge wins, and the ball is attracted to the comb.

"positive," and if there is a deficiency, it is "negative." Ben Franklin originated these terms for electrical usage, and apparently he merely borrowed them from arithmetic. He also brought in the plus and minus signs (+ and -). If you are thinking about electrons now, and you want to point out electrons are negative and move in wires, wait a bit. We will get over some perplexing problems by sticking to Ben Franklin's system; it's worked more than two hundred years now.

Now let us examine **attraction**, which is a basic property of an electric charge. For an experiment, we used a simple instrument like that in Fig. 3-1, which is a small piece of foamed polystyrene suspended from 8 inches of thread in a bent coathanger wire frame. Picking up a handy piece of plastic (a screwdriver handle worked well during the lab work for this book) we examined it very closely. No electric charge crawling around there. We brought it very near the ball, but nothing happened. After rubbing the plastic on some fuzzy synthetic cloth, we examined it again. Perhaps it was a bit shinier; some dirt has rubbed off, but there was no electric glow or any other indication of a really significant change. But now, when we brought it near the ball, the ball moved. The ball was attracted to it! Why?

This simple experiment stumped the world's best minds, if they tried it, for more than two thousand years after its first appearance in the writings of Thales of Miletus. We have the benefit of those men's thinking, and it gives us a workable answer. We say the contact of the plastic with the cloth transferred the charge from or to the plastic. It doesn't matter which way the charge really moved, and this detail depends upon what materials you happened to pick up. Maybe you can find a way to learn which charge went where, but at this point the key idea is "charge." Now, let us suppose the rubbing transfers the charge to a comb. When the comb approaches the plastic ball the effect is that shown in Fig. 3-2. The charge residing on the plastic comb generates an electric field around the comb. As the comb approaches the plastic ball the electric field repels positive charge in the ball, leaving the side near the comb negative. Charges that are relocated and held in place like this in some object in an electric field are called "bound" charges.

The negative bound charge in the ball is urged toward the positively charged comb, and the positive bound charge is repelled. But the negative bound charge is nearer the comb, so its attraction is the winning influence. The net force tugs the ball toward the comb. If we could find a way to deposit an overall positive charge on the ball, then the positively charged comb would always repel it. To learn how this is done, and to discover many interesting related experiments, read further in a good basic college physics textbook.

Now we are particularly concerned about the electric field around the positively charged comb. What is this field? How far does it extend, and how strong is it? A physicist knows

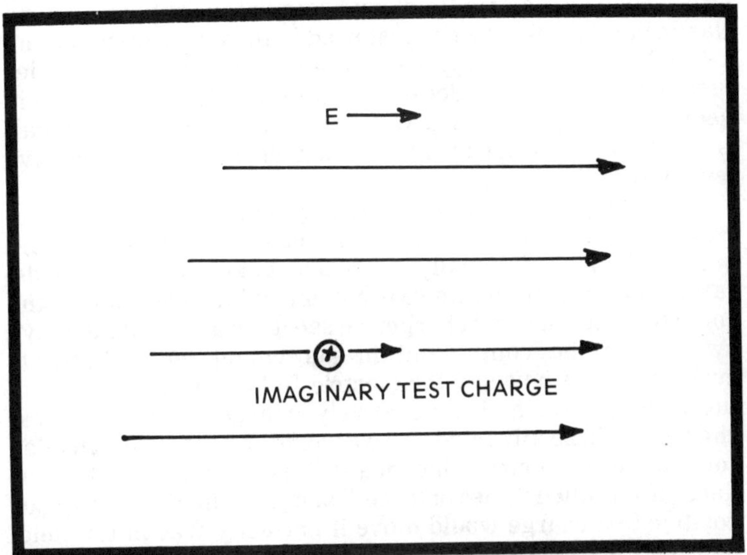

Fig. 3-3. Electric lines of force map the natural motion of a tiny positively charged object placed in the electric field.

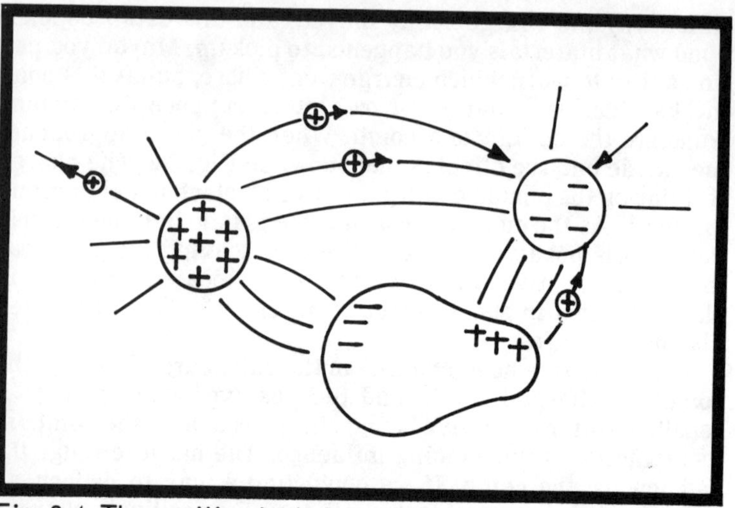

Fig. 3-4. The positive test charge convention makes it easy to describe complicated electric charge fields resulting from several charges near each other.

exactly how to describe electric fields. He thinks of fields as regions in space where some influence can be detected, and he describes them in terms of field strength and field direction. To describe an electric field, we begin with a tiny, imaginary plastic ball just like the one just used in the experiment, but on a smaller scale. We suppose this ball is so small and carries such a tiny charge it doesn't noticeably affect any electric field we might push it into. We suppose the charge on this ball is positive, and we speak of this little ball simply as a "positive test charge."

Bringing this positive test charge near the comb, we observe it is repelled, just as the ball would be repelled if it were positively charged. Usually, we do a test like this in our minds, rather than setting up an experiment. Whichever side of the comb we place our test charge, we see it always wants to move away from the comb, and in agreement with physicists everywhere we now say the electric field extends away from the comb. If we had a negatively charged comb, the test charge would be attracted and we would say the field extended toward the comb now. This idea is illustrated in Fig. 3-3. The lines, often called "lines of force," simply indicate the way our positive test charge would move if perfectly free in the field, and the closeness of the lines indicate field strength. Field strength is also indicated if we know how hard the field pushes against our test charge.

Now we are ready to picture a more complicated situation. See Fig. 3-4. Here we have a positively charged sphere, a negatively charged sphere, and a third sphere that is neutral but carries two bound charges. A test charge (several are shown) is attracted toward the negatively charged sphere and repelled by the positively charged sphere. Why are test charges attracted to one side but repelled by the other? The answer will help you understand how a strong electric charge developed in a thundercloud often results in a strong discharge to ground.

At this point we can say definitely why a charge flows in electric circuits. It is because there is an electric field inside the wire, and this field causes charged objects to move, transferring the charge from the positive terminal to the negative terminal along the wire. Unlike electric fields in space, the fields in the wire can be bent simply by bending the wire, and then the flowing charge goes where the wire goes. Why and how electricity flows in wires is basically as simple as that, as illustrated by the experiments with the plastic comb and ball. This explanation of electricity as mere electric charge, without any mention of electrons or currents, batteries, power supplies, and so on, comes from a small branch of physics known as "electrostatics." If we take that word apart, it comes out something like "fixed charges," and indeed that describes the kind of experiments we do to establish the appropriate details.

Circuits

If we leave the comb and other apparatus unused overnight, the next day we find the electric charges have all leaked back where they came from. The experiments won't work again until the comb is recharged. But what happens if we bring in some device that replaces the charge as it leaks off? Any such thing is known to the physicist or engineer as a "generator," even though it doesn't really generate a charge. As soon as we have a generator going, we find one terminal is positive (charge excess) and the other negative (charge deficiency). And we can always trace some route, usually one we planned, that a charge follows in its circulation from positive to negative generator terminals. A generator without any outside return is said to be "unloaded" or not loaded, since it cannot deliver any power.

But what happens if in some modern science machine we try very hard (accidentally perhaps) to operate a generator into an open circuit? A scheme for propelling spacecraft is

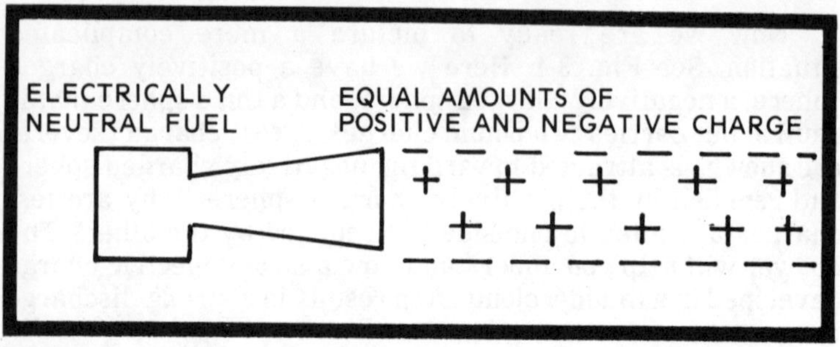

Fig. 3-5. An ion jet engine exhausts equal amounts of positive and negative charge.

near this. The ion propulsion engine shown in Fig. 3-5 works by shooting off a beam of particles very much like the exhaust of a chemical rocket engine. The particles are charged, so they can be accelerated down a very strong electric field inside the engine. The resulting reaction (Newton's third law) provides the propulsive force.

It turned out that an important problem in ion propulsion design was to guarantee the engine exhausted just as much positive charge as negative. That is, for every positive ion it kicked out it had to exhaust a corresponding negative ion or an electron. If this accounting didn't balance, the engine developed a terrific overall charge, its external electric field attracted the charged exhaust back to it again (completing an electric circuit) and, although it seemed to be working, its net thrust fell to zero. A modern ion jet engine is designed to exhaust an electrically neutral (average charge zero) plasma.

All routes that a charge follows resemble the "circuit" of Fig. 3-6. Older usages of this term "circuit" resemble "once around the route to where we started," and are still popular. For instance, we might say a bus makes one circuit of its route. The term serves well here because, like the bus route, the charge path must be complete from the generator's positive terminal around to the generator's negative terminal. One kind of commonly used circuit that seems incomplete, or "open," is the quarter-wave Marconi radio antenna and its relatives. Capacitors, too, look open. Actually, the circuit through antennas and capacitors is completed by the development of an electric field, which is accompanied by a "displacement current." We return to this in discussing capacitors.

Since charge always flows in closed circuits, we are in a position to see what electrostatics and test charges have to do with real circuits. We use these ideas to develop a plain picture

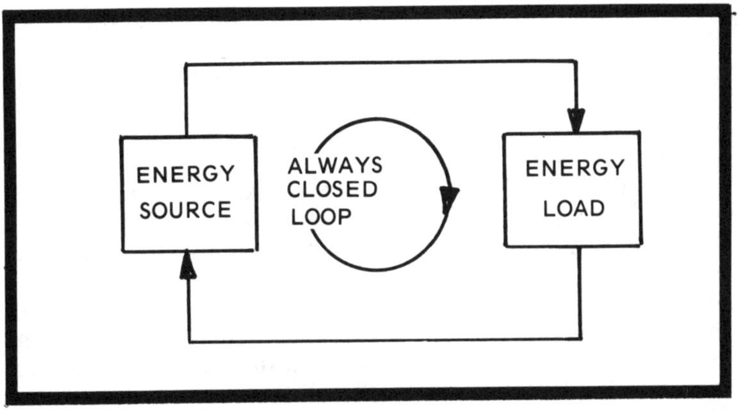

Fig. 3-6. An electric circuit means something like a "circuit of a race track." There must be a complete closed route from its beginning to its end—at its beginning.

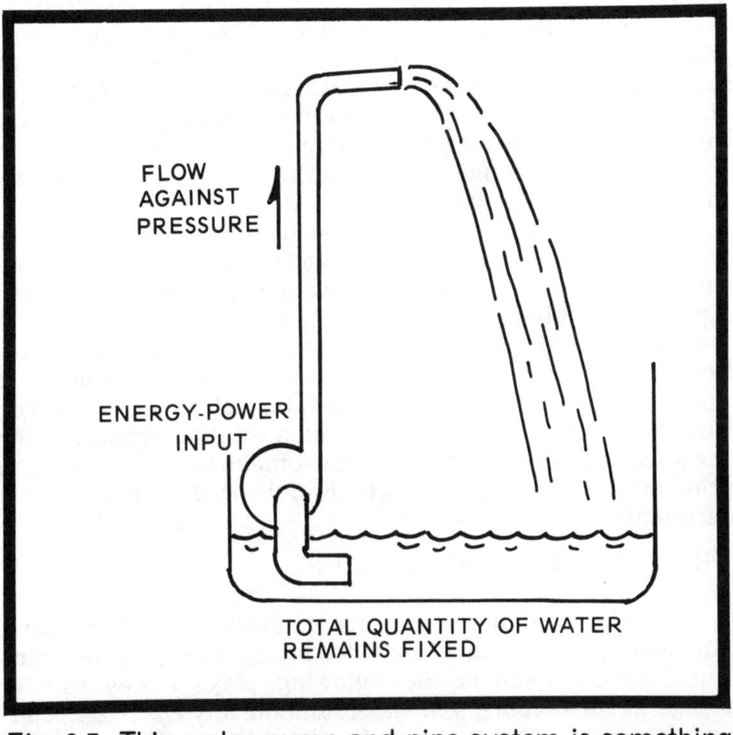

Fig. 3-7. This water pump and pipe system is something like an electric circuit. Study circuits until they develop their own identity; this idea will help you get started.

of what circuits are as seen by the charge flowing in them. See Fig. 3-7. Here we have a pump that pushes electric charge uphill, like water. The water is pumped **against gravity** to the top of a tube, falls to the bottom, and is pumped up to the top again. If we view the water as electric charge, we get a useful analogy to electric circuits.

The amount of water in the pipe circuit is fixed, and the water is motionless until we turn on the pump. Similarly, an electric circuit contains charge all ready to move, but nothing noticeable happens until we turn on the electric generator. The generator doesn't manufacture charge, it merely pushes charge. A certain water pressure is needed to get things moving—no pressure, no flow. If the water flows, then there is pressure. In electric circuits we call this pressure "voltage." We measure pressure across our pipe circuit as we measure voltage across an electric circuit. The greater the pressure or voltage, the greater the flow.

A definite amount of water flows, measured in gallons per second, perhaps. The electric charge flow is measured in coulombs per second. This is usually called "amperes," and since amperage or current is flow of charge, it isn't correct to speak of "flow of current." This works out to "flow of flow of charge," but many people use the term anyway. At any rate the greater the pressure the greater the flow, and to measure the flow we usually have to stick something into the circuit so the flow passes through it.

Finally, the amount of work done (energy spent) pushing water around the pipe circuit depends upon the pressure and the quantity pushed. This work is called "power" in electrical circuits, and it depends upon voltage and amperage jointly. If we turned off our pipe circuit we could pump it up to a certain pressure, seal it, and the pressure would remain without doing any more work. Similarly, a generator delivering no charge also does no work, although if it has a rotating armature, for instance, whatever is driving it must make up for mechanical friction even though no power is delivered at the electric terminals.

THE ELECTRICAL QUANTITIES

Six electrical quantities are fundamentally important, and adding one valuable basic idea brings the list up to a nice round seven. Each of the following plays a key role in designing, describing, and understanding any electric circuit.

Volt—electric charge pressure
Ampere—electric charge flow

Watt—electric energy or work
Ohm—friction opposing electric flow
Farad—charge-storing capacity
Henry—pressure-storing capacity
Duality—an important natural symmetry or balance

How these are applied to real circuits will clear up with practice. They usually appear as meter readings, or the results of calculations. "Duality" helps understand many kinds of circuits, particularly some kinds of logic circuits, and it will enable you to devise new ones easily.

Volt

A charge, anywhere, develops an electric field. A small charge develops a weak field; a large, a strong field. Two opposite charges develop a stronger field between them as they are brought closer together. All this follows from the definition of a field as a region where a test charge feels an electric force. But we cannot check our electronic circuits with little test charges, since we usually cannot get inside them.

Instead, we measure the potential—work-doing capacity—set up by the fields the charge would experience if we tried a test with it. We get an overall field strength in volts, where 1 volt is the potential rise that requires 1 newtonmeter (joule) of work to push 1 coulomb against it. But we don't even imagine doing this kind of experiment in practical circuit design or lab work. We simply clip on a simple test instrument, usually a multimeter or a very sensitive voltmeter which reads directly, to give us a voltage between two points.

Ampere

This is the unit of charge flow. It closely resembles "quarts per second," but for us it's coulombs per second. If the charge is moved by electron flow, that comes out 6.2×10^{18} electrons per second per ampere, and it's interesting to estimate how much actual mass is moving. Suppose we have a special ammeter-like instrument that adds up the weight of all the electrons that move into it. We connect this to a 100 ampere-hour storage battery and (gradually) discharge the battery. When the battery is run down, we discover the electrons making up our 100 ampere-hours (3.6×10^5 coulombs) weigh 2 milligrams. Since the electrons are merely transferred from one side of the battery to the other, the battery's overall weight remains the same.

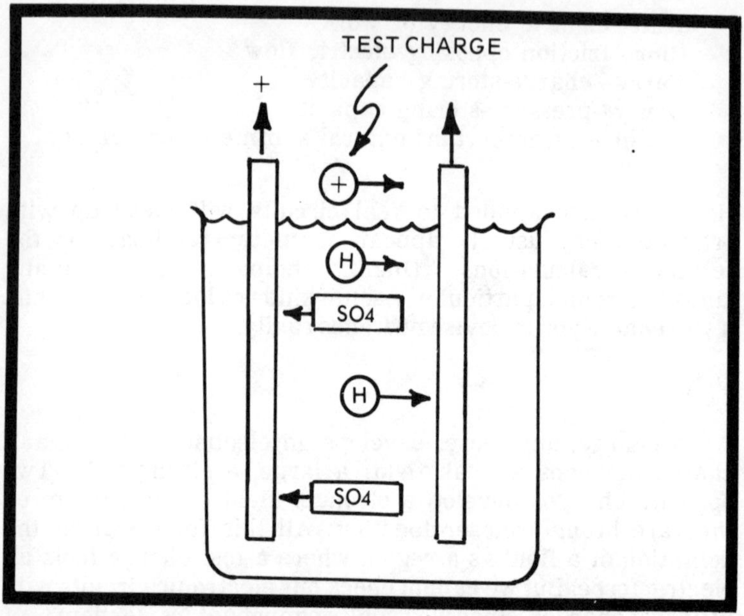

Fig. 3-8. Hydrogen and sulphate ions move in opposite directions in an electrolyte, but the conventional electric field extends from plus to minus. Hydrogen ions move with the field; sulphate ions move against it.

But charge can be carried by positive ions, too. Suppose we are electroplating copper, and we power our bath from our 100 ampere-hour battery again. It turns out our battery plates are out 119 grams of copper. How can 2 milligrams of electrons move all that copper? Simply because a copper ion weighs very much more than an electron. Michael Faraday pioneered the field of electrochemistry in the early 1800s.

We can now face the question, "Which way does current flow?" This question is poorly put to begin with, since current doesn't flow. Charge flows, and the current we measure in a circuit is the rate at which charge is flowing. And charge flows from plus to minus, since we built our system that way. Yet that reply seems unfinished. To improve it let's have a look at electrochemistry. Fig. 3-8 is an electrolytic cell with a battery connected to it, maintaining an electric field across its electrodes. Now, when we pour some electrolyte into the water the electrolyte dissociates into positively charged hydrogen ions and negatively charged sulphate ions. In response to the electric field between the electrodes, the ions **move in opposite**

directions. How do we simply describe this situation of electric flow with little particles moving both ways?

We use the concept of electric charge for this. We say that whatever happens, physically and chemically, charge flows from plus to minus. That settled, we now look into the circuit and see that charge is moved in wire circuits by electrons traveling from minus to plus. But in the electrolytic solution it is moved by hydrogen ions moving plus to minus **and** sulphate ions moving minus to plus (two H ions for each SO_4 ion; each type carrying half the charge transferred through the electrolyte). If we had some semiconductor material in the circuit, charge would be moved by electrons traveling minus to plus in N-type material, or holes moving plus to minus in P-type material. The charge concept unravels all these details and enables us to think about the circuit without accounting for everything that happens inside. It is one of the really basic ideas of physics and electronics—and it works.

Watt

The watt is the basic unit of **energy flow.** The basic unit of energy in circuits is the joule; 4.19 joules will warm 1 gram of water 1 degree Centigrade. A flow rate of 1 joule per second is 1 watt. Sometimes in electronics we work in joules; for instance, a 1000-microfarad capacitor charged to 10 volts stores .05 joules and a capacitor discharge photo flash unit could store over 10 joules. But usually we are concerned with steady flow, like that in a resistor which gets warm and stays warm.

A man talking as loudly as possible, if he keeps it up a while, puts out about 1 milliwatt (one thousandth watt) of sound power. We measure motors, hi-fi sets, fluorescent and incandescent lamps according to their performance in watts input or output. Powers under 1 watt or over 1000 watts tend to be rare. A well operated 1-watt transmitter can be heard at 100 miles quite easily, usually. The Mariner 4 that photographed Mars was later heard from the other side of the sun, 190 million miles away. It has a 10-watt transmitter.

For any DC electric circuit, watts consumed equals volts across the circuit times amperes through it. If we apply 8 volts to an 8-ohm loudspeaker, then 1 ampere flows through the speaker, dissipating 8 watts as heat and sound power. If a man speaking loudly generates 1 milliwatt, then the loudspeaker absorbing 8 watts should be very loud indeed. A 1-ohm resistor looking into 3 volts carries 3 amperes and dissipates 9 watts. If it is a small resistor it can get very hot. Most electronic components are rated in watts, in addition to other electrical properties. Resistors are supplied in one-quarter, one-half, one-, two-, and larger wattage capacity. This relates to how hot they get in operation. As a good circuit designer, you will work out designs that don't overheat components by choosing appropriate voltages and currents and by selecting components rated for greater loads then they will actually see.

Ohm

We know an electric current warms the conductor that carries it. Why is this? The electric charge flowing through the conductor is carried by an actual stream of particles; in a length of wire or a resistor these are electrons. The electrons cannot move freely in the wire, and warm it much as you can warm your hands by friction. The wire's internal resistance to electron flow is called resistance. Resistance is measured in ohms.

Since resistance is basically a mechanical property, it is stable and reliable. A conductor's actual resistance is the ratio of voltage across it to current through it. If our loudspeaker, just mentioned under watts, is passing 1 ampere when we apply 8 volts across its terminals, its resistance is 8 ohms. A no. 49 pilot lamp that passes 60 milliamperes at 2 volts has a working resistance of 33 ohms. (We will be using these lamps later, as indicators.) Friction or resistance in mechanical devices is usually a nuisance, and we try to minimize it. There are exceptions, such as a car's brakes or shock absorbers. But resistance in electronics is of central importance. All electronic circuits of any practical utility will contain resistors—maybe tens, hundreds or (in a computer) hundred of thousands of them.

Farad

When we rubbed our test comb across a piece of synthetic cloth and then brought it near a lightweight plastic ball, we observed an effect based upon a stationary electric charge generating an electric field. If we simply stood there and repeated the experiment 5 minutes later, we'd see the same thing happen again. In other words, electric charge can be fixed in place, or stored. The electronic component that does this is the capacitor. If we assemble the circuit in Fig. 3-9, we discover that charge flows into the capacitor, causing the lamp to flash, but the lamp quickly goes out. Then we can discharge the capacitor through the lamp, but again the lamp stays lighted only briefly. The capacitor can absorb and store charge, and release it again. This is something like our comb, which also stores charge and runs down after a while.

The capacitor's charge-storing capability is called "capacitance." Since charge is measured in coulombs we define capacitance in terms of a coulomb stored at some pressure; 1 farad stores 1 coulomb of charge at a pressure of 1 volt. Since the farad is too large for practical purposes (you can assemble a 1-farad capacitor from surplus electrolytics, for its novelty value) the familiar electronics basic capacitance unit is the microfarad, and we often see nanofarads and picofarads.

Timing systems are often based upon capacitor-charging circuits. If an ampere flowing 1 second charges a 1-farad capacitor to 1 volt, then a microampere-second will fill 1 microfarad to 1 volt also. Now, if we have a circuit that fires and discharges the capacitor when its voltage reaches 5 volts, and then resets, what will happen? The capacitor will fill up to 5 volts in 5 seconds, the circuit will fire and reset, the capacitor will recharge, and this cycle will continue indefinitely. This is one way electronic circuits can mark off units of time.

Capacitors are very much more interesting components than resistors (so are inductors) and their properties need more discussion. The similarity between a charged capacitor and a battery is very striking. A shorted capacitor will deliver tremendous amperage for a brief time, releasing the charge stored in it. This charge can be built up gradually by a low-

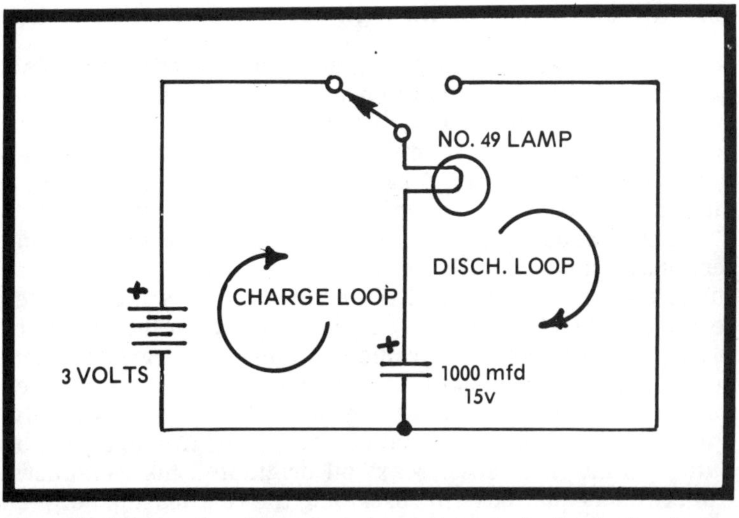

Fig. 3-9. The capacitor absorbs a charge to flash the lamp and releases it to flash the lamp again.

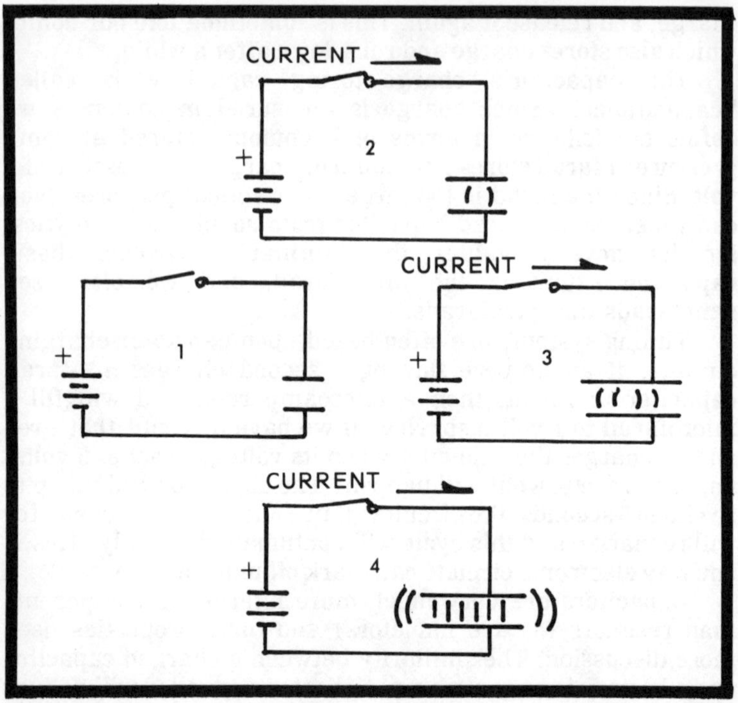

Fig. 3-10. An electric field develops in the space between a capacitor's plates. As this field develops, an electric current seems to flow through space.

power circuit, which then feeds a high-power circuit, such as a photoflash bulb or a jet ignition system. There is also a close similarity between capacitors and radio antennas. A circuit containing a capacitor seems to be broken, yet charge is seen to move in such a circuit. A radio antenna circuit can absorb a large radio-frequency current. Why is this? See Fig. 3-10.

What happens in each case is that charge flows in one terminal, an electric field develops in the open space where the circuit seems to be broken, and charge flows out the other terminal. The electric field developed between the capacitor plates or between the radio antenna and ground acts something like a length of electrical conductor—it even develops a magnetic field as it grows. This current-in-empty-space, called a "displacement current," was first analyzed by James Maxwell in 1865. Maxwell originated his well-known "Maxwell's equations" in describing the relationship between electric and magnetic fields, and now his work is seen as the foundation of radio electronics and communication technology. We will see little of displacement current in this book (except to remember it appears between the plates of every capacitor we use).

Henry

The original Henry of the electrical world was Joseph Henry, who built the world's first electric motor in Albany, New York. He also experimented with electromagnets, and discovered the basics of telegraph systems that Samuel Finley Breese Morse developed into a working telegraph technology (without acknowledging Henry's work). Today the henry is our standard unit of inductance.

If capacitance stores energy in amp-seconds per volt, might there be something that stores energy in volt-seconds per amp? It is the inductor. A 1-henry inductor will develop a 1-ampere current in 1 second if the voltage across its terminals is held at 1 volt. The capacitor delivers any possible current trying to fix the voltage across its terminals; the inductor delivers any possible voltage trying to fix the current through its terminals. The experiment in Fig. 3-11 shows this. Here we have a radio-type inductance of perhaps 10 to 30 henries. We connect one terminal to a battery. Then we place the other terminal on the other side of the battery briefly, and lift it off. We see a spark or arc of much greater voltage than the battery can deliver. This is the inductor's doing.

It means, if we use inductors in transistor circuits, that we must be careful the circuit does not ask the inductor current to change too quickly. For instance, suppose a transistor turns

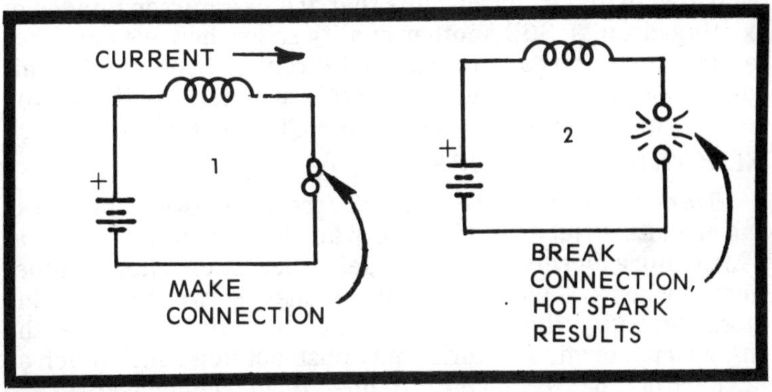

Fig. 3-11. When an inductive circuit is broken, high voltages develop. This is easy to demonstrate with a battery and an iron-core choke. An "inductive kick" can destroy transistors.

off the current through an inductor; the inductor then delivers a high voltage trying to maintain the current, and if the inductor wins the transistor is destroyed. Correct circuit design manages this danger—or uses it as a simple way to develop high voltage not otherwise available from the circuit.

From a mathematical view, inductors can do all the things capacitors can, but this doesn't work out in real circuit design. A capacitor is typically a number of large plates electrically wired in parallel. Electrical resistance in such a mechanical structure tends to be small. But an inductor can be made only of many turns of wire, which means a considerable length of wire. This arrangement is made to order for resistance, whose effect reduces the usefulness of the inductance. The difference between inductors and capacitors in terms of their electrical effectiveness becomes less important in high-frequency and pulse work but we won't be using capacitor-dual circuits in this book.

Duality

Here we make one of our closer approaches, for this book, to a more abstract part of electrical science. There are certain basic symmetries we can find in nature, if we look closely. Some of these have turned up in preceding paragraphs. We found capacitance was measured in amp-seconds per volt, and inductance in volt-seconds per amp. Of course, both can be measured in many other terms, too. We will find another duality in transistor circuits, where we have our choice of using PNP or NPN transistors, and we can reverse supply polarities to make new circuits that are like mirror images of existing circuits. Still another duality shows how we can often replace series circuits by parallel circuits. Look out for this curious same-yet-different key to things; later it will help you design and understand more challenging, useful circuits.

ELECTRICAL NETWORKS

To begin with, there is no difference between networks and circuits. "Network" is simply an electric circuit looked at a little more closely than we usually look at circuits. In other words, when we use this term we suggest we are thinking more carefully than when we say "circuit." In time the engineering term "network" may push out "circuit," much as "capacitor" has replaced the older term "condenser."

We are interested in networks because we know certain things that serve as a foot-in-the-door to understanding how designs work. In order to develop a picture of what happens inside a circuit we must begin somewhere. Usually an idea from or based upon the following network theory basics provides the necessary opening. Then past experience and knowledge, guessing, calculation and measurement fill out our picture of detailed voltages, currents, signals, and sequence of events.

Electric Charge Is Conserved

This fact was mentioned earlier, but it deserves repeating. "Conserved" means the circuit does not invent electric charge or drop it by the wayside. If our circuit contains one generator, for instance, and we can account for all the charge that passes through this generator, then we have finished the job of understanding the circuit. We also know if a circuit has no generator in it, then no charge moves; if a circuit **seems** to have no generator and yet somehow we detect charge moving in it we must look again; we've made a mistake. Until we find the generator we do not understand the circuit.

Node Theorem

This is also known as Kirchoff's First Law. A "node" is any point where two, three or more circuit branches join. We

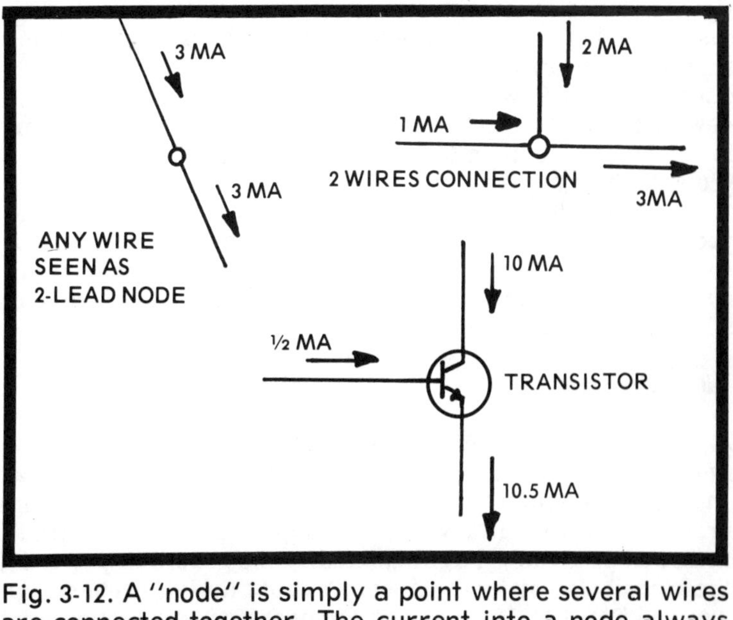

Fig. 3-12. A "node" is simply a point where several wires are connected together. The current into a node always equals the current out of it.

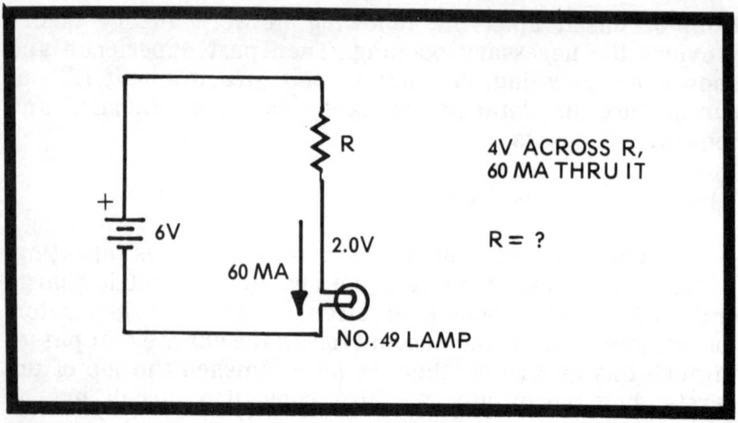

Fig. 3-13. The voltage across the loads in a circuit loop equals the supply voltage. Here, we are designing a circuit that reduces 6 volts DC to the 2 volts at 60 ma needed by a No. 49 incandescent lamp.

could view a point in a single piece of wire as a node, if we wished; a joint of two branches. Actually, that is too elementary to be useful. See Fig. 3-12. It's more practical to view the junction of three or more leads as a node. In all such cases our charge conservation theorem tells us that as much charge flows out of this node as into it. If we measure 20 microamperes flowing into it, for instance, and we find 10 flowing out one branch, 5 out another, and that the third seems to be getting 10 microamps, then we have come upon a problem. Five microamps must be accounted for. We could be talking here about a transistor base bias circuit. This unexpected 5 microamps might be a collector-base leakage. Also, it could be reasonable to wonder, if we were working on a circuit board, if the board might be leaking the extra current from an adjacent conductor.

Loop Theorem

Since all charge flows in circuits from generator around to generator again, all circuits must be "loops." The sum of all the voltages around a loop must be zero. What this means practically is that all the voltages appearing across a circuit must add up to the voltage supplied by the generator powering the circuit, for instance. A very simple design problem appears in Fig. 3-13. Here we want to operate a 2-volt pilot lamp from a 6-volt battery. Our loop theorem says the sum of the voltages across R1 and the no. 49 pilot lamp must equal the 6 volts across the battery. If there is 2 volts across the lamp, then 4 volts appears across R1. If we knew how much current was supposed to flow, we could choose a value for R1. But this no. 49 lamp is mentioned elsewhere in the book, since it appears in several circuits. Its rated current is 60 ma. If R1 has 4 volts across it when 60 ma is passing through it, then its resistance must be 4 volts divided by .06 amperes or 66.7 ohms. If you remember your electronics "magic numbers" you instantly hit upon 68 ohms as the real resistor value nearest this computed result.

We did something here you will use often when designing circuits. Looking back, you will discover a certain point where the thinking changed from describing a problem to supposing the circuit was finished and working. The revised problem looked like a detective's assignment to discover something, given several clues. This little change in emphasis, advancing little by little upon parts of the problem, often solves tough problems surprisingly easily. Now, one detail remains. Good design practice is to keep resistor power dissipation at or under 60 percent of the manufacturer's ratings. A 1-watt resistor should not exceed 0.6 watts dissipation in a real circuit, and so on. The limit can be pushed up if the resistor is in a cool, well-ventilated place. What is the smallest resistor usable here? A quarter watt? A half watt?

Maxwell's Method

A simple, popular circuit that cannot be analyzed easily is the wheatstone bridge, a versatile instrument circuit. Various analytical methods appear in textbooks; one is not so commonly mentioned, but it is well worth looking for when you want to learn more about analyzing circuits. This is "Maxwell's Method." It's not described here, but you can find it in an electronics textbook. Try Brophy's "Basic Electronics For Scientists," p. 23, where it appears under "Wheatstone Bridge." It's not indexed properly in that book. An evening's work understanding Maxwell's Method will teach you more about network analysis than any amount of careful applications of Kirchoff's node and loop theorems.

BATTERIES AND POWER SUPPLIES

In modern electronics we meet two basic types of charge "pushers," or generators. These are batteries and power supplies. Batteries are fueled during assembly with chemicals whose combination provides the energy available at their terminals, and some of these can be "recharged." That is, the energy they have supplied to an outside circuit can be pushed back in again. And power supplies are circuits that absorb energy from other sources, such as an alternating-current wall outlet, and feed it out in a form usable by electronic circuits. There is no reason to absolutely prefer either one over the other.

Batteries are convenient, portable, simple, and rarely (but sometimes!) introduce noise into a circuit. Power supplies offer practically unlimited power, both over time and in magnitude. But batteries do run down, and power supplies fail if their power sources fail; also, power supplies are either fairly expensive or their construction uses up time usually better spent on circuits. Yet battery power is costly compared to power from local AC circuits, and a well built supply should run many years without attention. The wise electronics worker chooses to operate a given circuit or system from battery or power supply, or some combination of both, upon the merits of the particular situation. It is not always an easy choice.

The circuits in this book were designed to work from 6- or 12-volt batteries or supplies. Such sources are available almost anywhere, and the circuits can be taken into the field and operated from a car's electrical system. Other supply voltages could be chosen, and 9 volts is a favorite value. Transistor circuits can be designed to operate at 3 or even 1.5 volts, with some efficiency loss. This is made up by a few added components, and such circuits have a very definite utility and novelty value.

How Batteries Work

All batteries work about the same way, only the details vary. Batteries burn fuel. In fact, battery chemistry is so basically like that of flame or other chemical reaction involving oxygen that chemists lump battery chemistry and actual combustion under the same label: oxidation reactions. The basic process that pumps charge through batteries depends upon a mechanical arrangement like that shown in Fig. 3-14.

Here, we have fuel, oxidizer, and a leaky wall between that prevents extremely rapid mixing. This assembly is sandwiched between some kind of electric terminal plates or electrodes for carrying charge out the positive terminal and in the negative. The leaky wall may be a boundary between two solutions that mix poorly, as in the Daniel cell, a wet, porous insulator as in a storage battery, or a layer of moist chemicals as in the so-called "dry cell" (which contains a surprising amount of moisture, if it leaks corrosive goo over your valuable electronic circuits).

When the cell is assembled, ionized atoms or molecules of fuel and oxidizer pass into the barrier region and a state of tension—an electric field—develops from terminal to terminal across the cell. (Which way does this field go?) If no charge flows, as before the cell is connected to an electric circuit, the field stops the chemical processes that developed it. The situation stabilizes at a standoff with excess charge at the positive terminal and a charge deficiency at the negative terminal.

Now let us add an external circuit that absorbs electric energy. The field inside the cell is weakened, chemical processes begin again, and charge flows. If we increase the load the field is weakened further, the chemicals combine faster, and more charge flows. The chemical combination or burning rate is sharply dependent upon the electric field inside the cell, so a small field reduction is balanced by a strong increase in chemical reaction rate. The cell seems to be trying to deliver whatever current is needed to hold its inside electric field constant against the outside load. All cells work this way, but the chemical details are usually complex. One exception is the fuel cells that powered the Apollo space ships. These burned hydrogen and oxygen into water.

If an oxidation reaction (like burning) goes on inside a cell, what happens to the waste products? Usually, they stay

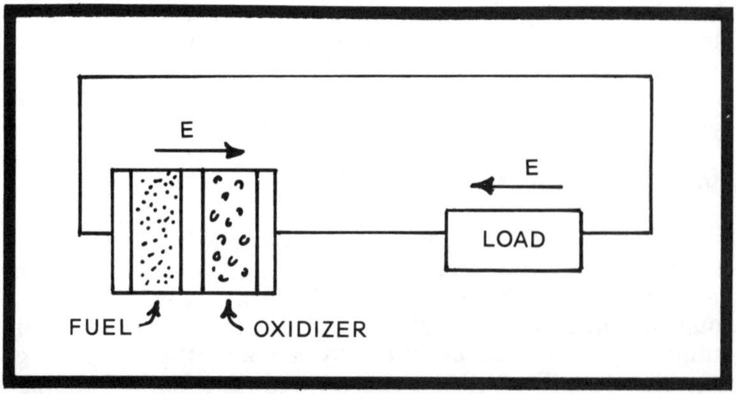

Fig. 3-14. All batteries and fuel cells work according to this basic scheme. Controlled mixing of fuel and oxidizer across a leaky wall develops an electric field.

right in the cell (waste water from the Apollo fuel cells served as drinking water). Under heavy loads waste may reduce cell efficiency because it cannot move out of the way fast enough. And as the cell ages with use, a gradual overall buildup of waste products permanently reduces efficiency.

Why can't the burning chemistry be reversed by pushing charge through the cell against normal flow? Sometimes it can, and cells designed to work this way are known as "secondary" cells. A lead-acid car battery, for instance, is a secondary cell. Nickel-cadmium cells are also secondary cells, and certain alkaline manganese cells are designed for rechargeable operation (most are not). Cells not designed for recharging can be revitalized to some extent and there are several "rechargers" on the market. Suppliers who offer these things to the public help only themselves.

Preferred Cells And Batteries

If we rate certain cells and batteries according to the cost per kilowatt-hour of energy purchased in them, we learn two important facts. First, the cheapest battery power is much more expensive than AC-mains power. Secondly, some cells and batteries supply much more costly power than others. The popular 2U6, a tiny 9-volt battery, is extremely expensive. The large and commonly used D cell is probably the least costly energy source available anywhere, unless you make your own (a common practice once; maybe you should think about that). A special point favoring D cells is that they are available almost anywhere, simplifying logistics for those engaged in scientific work somewhere away from things. And the price of energy from car batteries or true rechargeables isn't mentioned here since it depends upon how these sources are used, but it tends to be well below that of D cells. Fig. 3-15 lists the advantages and disadvantages of various battery types.

How Power Supplies Work

A power supply is basically a simple circuit consisting of a transformer that converts the high AC line voltage to a lower voltage; a circuit that changes the low-voltage AC into a very rough DC, and a circuit that improves the raw DC to electronic standards. This last circuit may be as little as a large capacitor that fills in the "hollows" and absorbs the "bumps" in the power supply output to an elaborate system that holds voltage accurately constant. It may even serve as a watchman who turns off if the load or its own circuits get into trouble.

CARBON-ZINC (LECLANCHE)	LEAST EXPENSIVE, AVAILABLE ANYWHERE. FAIR PERFORMANCE; WEAK AT LOW TEMPERATURES.
ALKALINE MANGANESE	MODERATELY EXPENSIVE, HIGH PERFORMANCE. EXCELLENT AT BELOW-FREEZING TEMPERATURES.
MERCURY	MAXIMUM ENERGY PER SIZE AND WEIGHT. EXPENSIVE. USELESS NEAR AND BELOW FREEZING.
NICKEL CADMIUM	MODERATELY EXPENSIVE CELLS, RECHARGEABLE HUNDREDS OF TIMES FOR CHEAPER BATTERY POWER THAN ANY OTHER TYPE. CAN DELIVER HIGH CURRENTS.
LEAD-ACID	AUTO BATTERIES STORE LARGE ENERGIES, CAN DELIVER HUNDREDS TO THOUSANDS OF AMPERES. NEED WATER REPLENISHMENT, EMIT SMALL AMOUNTS OF ACID.

Fig. 3-15. The most common types of primary and secondary cells. The carbon-zinc types are least expensive, but alkaline manganese types are sometimes preferable for transistor circuits.

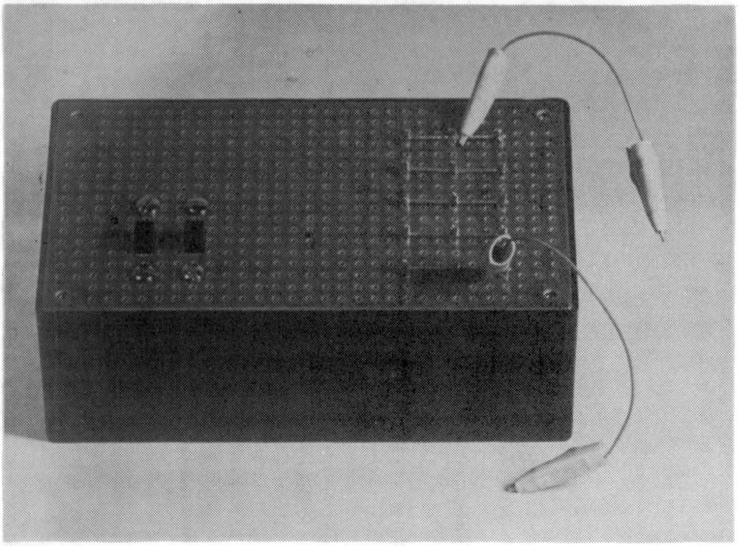

Fig. 3-16. This simple battery box is very useful in amateur electronics work. If new voltages are needed, different battery carriers are clipped in. The wire terminals have proven better than any connectors, using clip leads for connections.

Most power supplies for amateur work lie somewhere in between these two extremes.

A Plug For Batteries

Most circuits in this book were operated from batteries simply because batteries are nicer to work with than power supplies. The greater cost of battery power is unimportant since you use little of it. Batteries' wining points are convenience, flexibility and adaptability (batteries can be borrowed from one job and plugged in to another), safety, and freedom from noise carried in along AC lines.

A simple battery power supply, the one used for most work in writing this book, appears in Fig. 3-16. It is simply a box with two switches wired as shown in Fig. 3-17. There are no jacks or plugs because clip leads work out better. This reduces the cost a few cents too. The batteries inside are good grade alkaline manganese pencells in convenient plastic holders. To change voltage, one holder (say four cells for 6 volts) is unclipped and another (say six cells for 9 volts) replaces it. This arrangement is simple, convenient, inexpensive, and it has proven better than any other DC supply system I've used, even if it doesn't look very scientific.

THEVENIN'S AND NORTON'S THEOREMS—THE GREAT SIMPLIFIERS

Real electric circuits that supply power vary from simple batteries to elaborate circuits and systems. They look very different, but that apparent divergence is not as total as it may look. For instance, a high fidelity rig that feeds an audio signal into a loudspeaker is, from the speaker's point of view, quite like a battery that feeds a lamp, as the lamp sees it. It makes no difference that the one is an expensive, elaborate electronic system and the other a small can of watery chemicals soaked into some powdery or muddy ones. These two kinds of things are somehow, importantly, alike.

Two such likenesses, pared down to essentials, are known as "Thevenin's Theorem" and "Norton's Theorem." These theorems (theorem: a well-established idea) give us a way to represent complex circuits in terms of their essential performance, all other details being omitted. The resulting

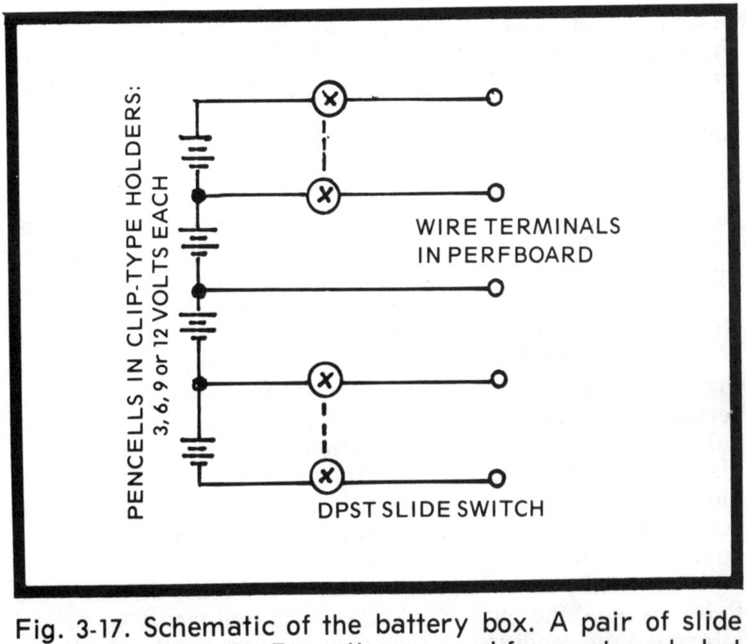

Fig. 3-17. Schematic of the battery box. A pair of slide switches work well. Pencells are good for most work, but heavy current applications may require D cells.

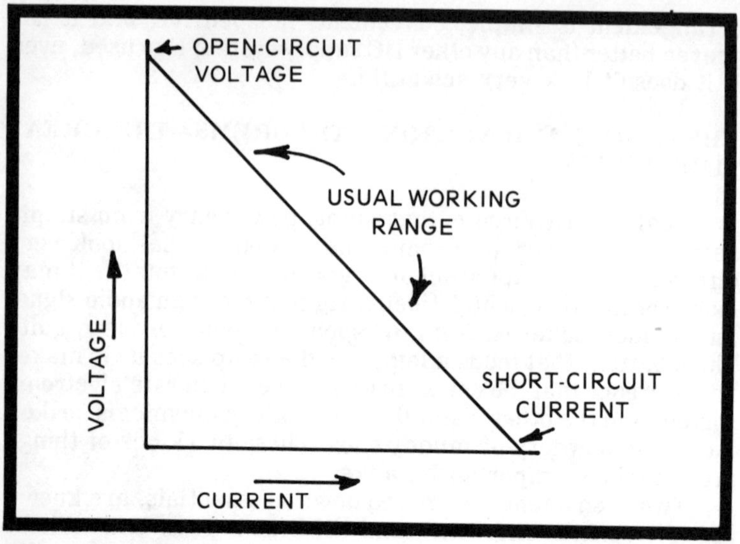

Fig. 3-18. A load line gives a battery or power supply output voltage for any current load from open circuit to short circuit.

descriptions are, to somebody who has been trying to work without them, unbelievably simple. We begin by looking at the way power supply or signal circuit voltages drop under load.

Load Lines

Practical experience says that if we test a battery, a signal source, or any other AC or DC charge supply system, we find it delivers a higher voltage into an open circuit than into any other kind of load. (AC circuits seem to be exceptions, but they're really not.) And if we feed the output into a short circuit we find a higher current than exists with any other kind of load. The load dissipates no power in either case, since in the first there is no current through the open circuit and in the second there is no voltage across the short. Since power equals voltage times amperage, if either is zero no power is dissipated in the load whatever the other may be. Then, if we plot the open-circuit voltage and the short-circuit current, connect these with a straight line and examine the result, what do we have? See Fig. 3-18. The line is a "load line."

There isn't any compelling reason to expect this load line to tell us what the output voltage or current would be at any intermediate condition between open and short circuit extremes, but it turns out we can pretty accurately determine any intermediate voltage and current values if we know these extremes. Yet these extremes may harm the generator. For instance, if we short circuit a D cell to discover its maximum current capacity, we'll get a reading all right but the cell won't ever quite recover from the test. And if we shorted a powerful lead-acid secondary battery we'd be faced with additional problems of measuring thousand ampere currents. We replace hard tests with reasonable ones.

If tests at extreme conditions tell us something about what happens at moderate conditions, we can as well work the other way. We simply measure voltage and current at two easily managed values, not too similar. We plot the resulting points and extend our straight line to an estimated open-circuit voltage and short-circuit current. When working with chemical batteries, the open-circuit voltage is easily measured and serves as one point. Let us pick up the battery supply previously mentioned and work out its load line.

Open-circuit voltage across two of its terminals turns out to be 9.39 volts, evidently six 1.5-volt cells and these are AA carbon-zinc cells. Upon hooking an 82-ohm resistor across the terminals, output voltage drops to 8.55 volts. The resistor is taking 104 ma, according to Ohm's Law and a sliderule computation. If the terminal voltage dropped 0.84 volts at 104 ma, it would drop another 0.84 volts upon increasing the load another 104 ma; eventually it would come to zero. A simple calculation or accurate graphical construction gives us the load line of Fig. 3-19, with a short-circuit current of 1.16 am-

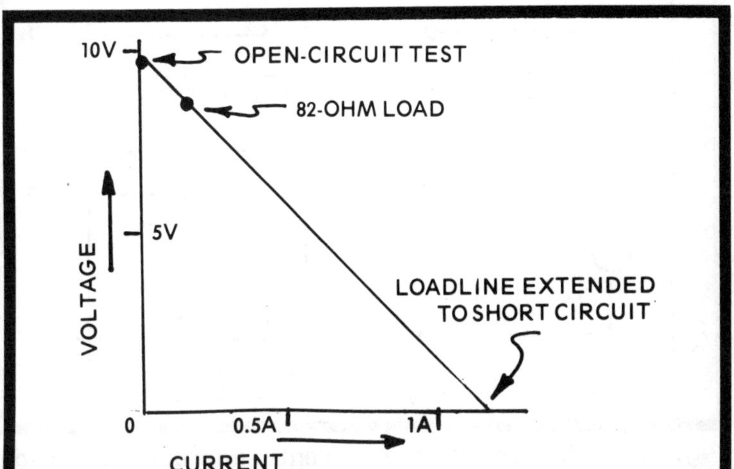

Fig. 3-19. This load line was developed from tests on batteries in a pencell battery supply.

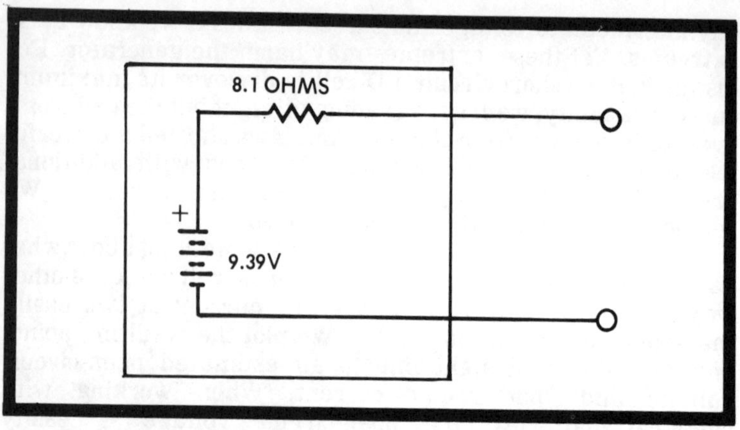

Fig. 3-20. An "equivalent circuit" that describes the battery box loadline by supposing that the box has a perfect battery inside, plus a register to cause a voltage drop under load.

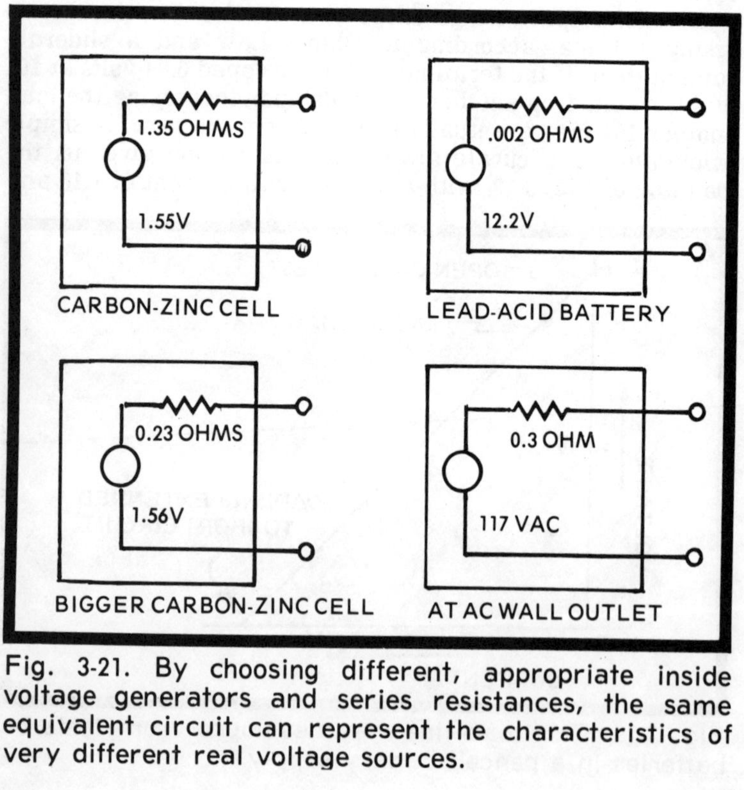

Fig. 3-21. By choosing different, appropriate inside voltage generators and series resistances, the same equivalent circuit can represent the characteristics of very different real voltage sources.

peres. We can now use this load line to estimate the terminal voltage we would find at any definite load current or to estimate the effect of pulsed loads upon output voltage. At very high currents we would see the actual voltage falling below the load line since chemicals in the battery cannot keep up with this out-of-specs load. But we try to operate batteries and all other components within reason, if not entirely within specs. This detail is reserved to complete our picture of things, but has little practical value.

Thevenin's Theorem

If all circuits that deliver a voltage output depending upon load act as shown in Figs. 3-18 and 3-19, why cannot they be all represented by the same general circuit? One circuit that fills the bill is Fig. 3-20. In this case it is a 9.39-volt battery in series with an 8.1-ohm resistor; these are enclosed in a box we cannot open. We suppose the battery voltage is unaffected by load, and the 8.1-ohm resistor is responsible for the voltage performance we found at the terminals. This gives us a concise picture of our batteries' properties called a "Thevenin equivalent" circuit. The same circuit, with only generator and equivalent resistance values changed, can illustrate the properties of many real circuits. Examples appear in Fig. 3-21.

Norton's Theorem

The load line in Fig. 3-19 could be produced by another kind of equivalent generator. Suppose we have a generator that always puts out a constant current, rather than a constant voltage. It would be happy delivering no power into a short circuit, just as the Thevenin generator likes to deliver no power into an open circuit. Now, as the connection across its terminals is changed to increasing resistances, the total current provided by the current generator remains fixed. But a decreasing part of it flows through the outside load as more and more is diverted to the generator's internal equivalent resistance. Finally, when an open circuit appears across the terminals, all the generator's current flows through the equivalent resistance, resulting in the observed open-circuit voltage. This outside performance is exactly like that of the Thevenin generator, but an entirely different (a dual) system is supposed to exist inside the dotted line. See Fig. 3-22.

What is the meaning of this? If we cannot distinguish, from outside the box, between a Thevenin and a Norton equivalent generator, why propose two of these? Because in

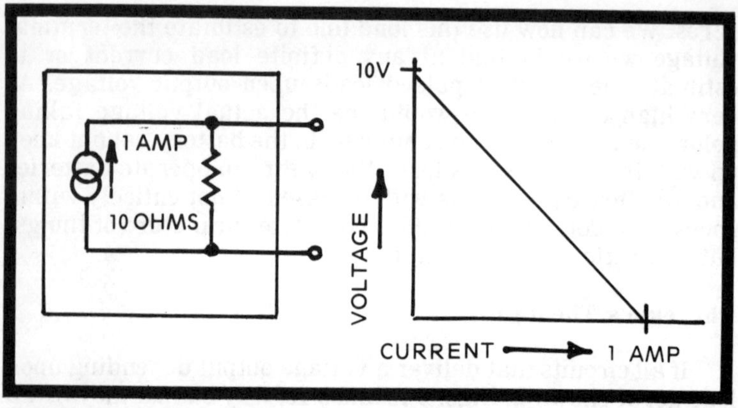

Fig. 3-22. A Norton equivalent circuit. This supposes the generator delivers a constant current rather than a constant voltage. What Thevenin equivalent would have the same loadline?

design work we are very interested in the assumed perfect generator's properties and the assumed equivalent resistance, but not always both at the same time. And our problem or general context may lead us to think about voltage so that we prefer the Thevenin equivalent, or when working in current terms we like the Norton equivalent. Many examples appear later; some very soon.

DIODES

Some people puzzle over a supposed difference between electricity and electronics and try to define it. Things aren't that simple. The one grew out of the other and observers can draw a line reasonably almost anywhere. A historian might put it at the time of Thales of Miletus in 2500 BC, since the effects Thales observed were due to the accumulation of electron-carried charges in various insulating materials. At the other extreme, a tough-minded engineer would assert that true electronics came in with integrated circuits. But all that's beside the point—systems analysis is the field today.

At any rate, the diode comes through clearly as the simplest electronic component (some disagree anyhow, citing "twistors" and "ovonics"). A diode resembles a piece of wire, but with differences. An electric charge applied to wire results in charge flow, whichever direction the field. But diodes are something else: they conduct charge one way but not the other. We begin with some semiconductor physics.

Semiconductors are crystalline or glassy materials that conduct electricity poorly. A pile of dirty carbon also conducts electricity poorly, but is neither crystalline nor a semiconductor. A windowpane is glassy but conducts electricity not at all; it isn't a semiconductor. A piece of quartz and a crystalline rock are not semiconductors, although they are crystalline. "Semiconductor," to the electrical engineer or the solid-state physicist, denotes a material that becomes a moderately good conductor if it is doped with a few parts per million or billion of certain materials. The doping falls into two general classes, the "N" for semiconductors that move charge in electron carriers and "P" for semiconductors that move charge in "holes" or missing electron carriers.

Let us look into a piece of N-type semiconductor. We find it has an orderly structure, which does not concern us; its important property is a lot of free charges available to move from one terminal to another. If we put a block of this material into an electric field, as shown in Fig. 3-23, the electrons concentrate on the side of the block nearest the positive side of the field. The longhand E and arrow indicate which way a positive test charge would move if we released it in the field. Now, if we replace this block with a P-type semiconductor block, we get a very different picture. The charge carriers in P-type material are "holes," places in the crystal structure where there ought to be an electron but there isn't. Missing-electrons or holes act like little positive charges equal in value to electrons, but they tend to the negative side of the block when we put it into an electric field (Fig. 3-24).

Since about 1948 it has been easy to make semiconductor blocks like Fig. 3-25. Here we have one single piece of crystal

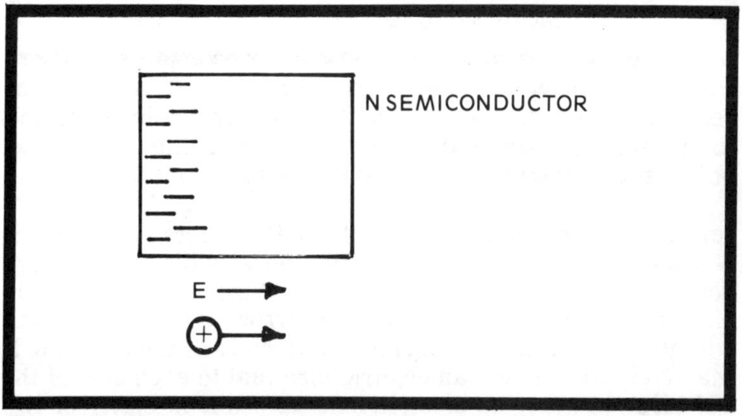

Fig. 3-23. The electrons in a block of N-type semiconductor all move toward the plus side of an electric field.

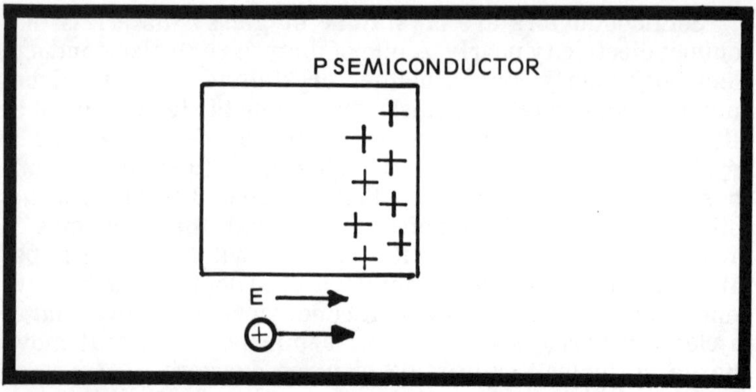

Fig. 3-24. The holes in a block of P-type semiconductor all move toward the minus side of an electric field.

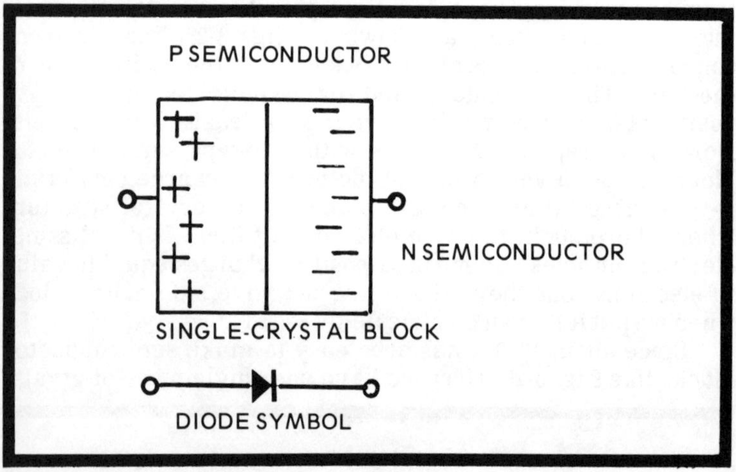

Fig. 3-25. If we make a PN junction in a piece of semiconductor crystal, a few electrons and holes combine to develop an electric field that pushes the remaining holes and electrons away from the junction.

semiconductor, but part of it contains P carriers and part of it N carriers. There is a physically observable, very sharp boundary between the two. Such structures are indicated in electronics circuits by a combined arrow-and-block symbol, with the arrow representing the P material and the block the N material. We can add an electric terminal to each side of the PN block to apply an electric field across it in either of two directions. These are termed "forward" and "reverse." Let us see what happens in each case.

In Fig. 3-26, with the battery's negative terminal connected to the P and its positive terminal to the N material, the electric field separates the charge carriers. The electrons are attracted toward the positive side and the holes toward the negative side. No P or N carriers approach the junction; therefore, no charge passes across it. There is no current and the PN junction is said to be "reverse biased." This leads to a useful fact if we bring in Kirchoff's loop theorem. The battery applies an electric field across the overall circuit. But how is the field distributed through the circuit? It is not distributed equally all around. Since no charge flows, the tiny resistance of the leads develops no voltage drop. Yet we know there must be some electric field somewhere if the voltage across the circuit equals that provided by the battery. If it's not in the leads, then it must be in the PN block; things must be rather tense in there.

The PN junction is something like one ten-thousandth inch thick, or less. If the 6-volt battery field appears across this distance, that works out to a potential gradient or steepness of 60,000 volts per inch! If fields like this appear in diodes, then the good circuit designer thinks about planning his circuits so the diodes aren't abused by too much of this stress. In real life, the experienced worker knows that diodes are lost this way. But sometimes a diode's reverse voltage limit is deliberately exceeded; this is discussed in Chapter 4.

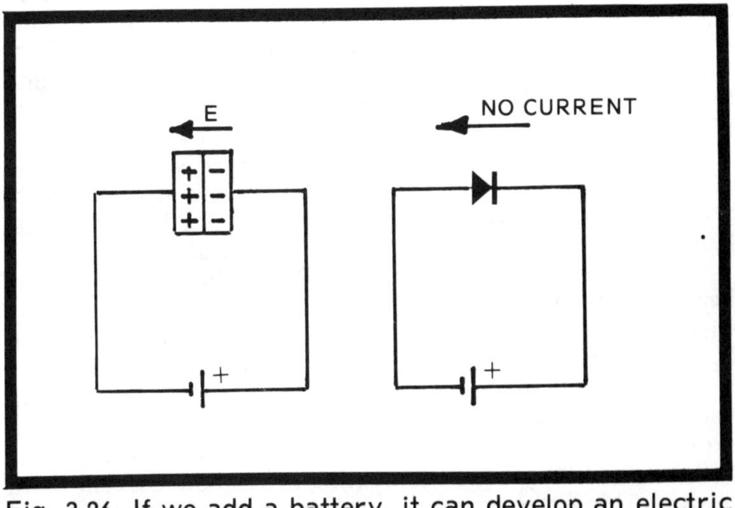

Fig. 3-26. If we add a battery, it can develop an electric field that further separates the electrons and holes. This diode is said to be "reverse biased."

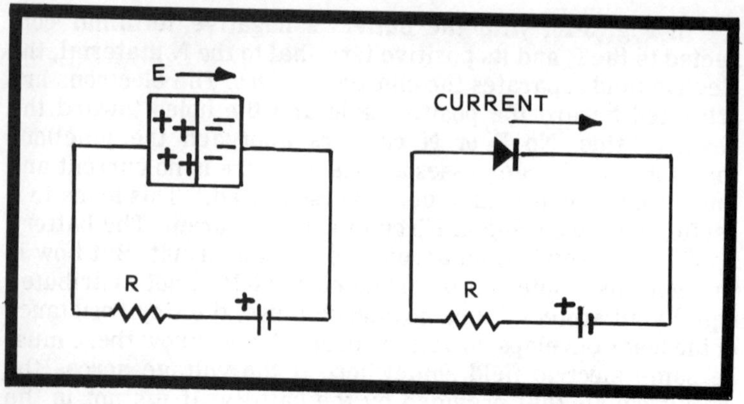

Fig. 3-27. The battery can also be connected so that its electric field overcomes the PN junction's natural reverse field and brings the electrons and holes together. Then, an electric current flows and must be limited by a series resistance outside the diode.

In Fig. 3-27, we reversed the battery and hooked it up again with a series resistor. The reason the resistor is there will emerge shortly. Looking at the PN block we see that, with the electric field reversed, both electrons and holes are moving toward the PN junction. The electrons move against the field and the holes along it (being positive) and where they meet, each falls into the other and they seem to vanish. This arithmetic works, since electrons flow in the N material from the wire connection; electrons flow out of the P material into the wire on the other side, leaving the holes in the P material to move toward the N material. It does seem kind of odd. So, although charge seems to vanish at the PN junction, the overall arithmetic is right. Charge moves; a current is observed. The diode is now said to be "forward biased": its behavior looks, from the outside, like a piece of wire.

Now, we come to the resistor R1. If the PN junction is forward biased and charge flow commences, what sets the current? The wire leads have a very low resistance. The P and N semiconductor conductor materials have moderate resistivity, but they are very small. Evidently, a forward-biased diode is a piece of wire so far as its current limiting properties are concerned (much more on this in Chapter 4) and it is necessary to add something to keep the current down in sight. This is the purpose of R1, which sees nearly the entire battery voltage when the diode is forward biased. If there is any doubt about this, tests with (inexpensive) unprotected diodes will establish they appear to fail instantly (it actually takes microseconds to seconds, depending upon details).

The circuits in Figs. 3-26 and 3-27 should be breadboarded and tested, preferably with close observation of voltage, current, and their distribution around the circuit. This experience illustrates the basic rules for diode circuit design: when a diode is reverse biased, the circuit must choose the voltage applied to the diode. And when a diode is forward biased the circuit must determine the **current** that passes through the diode. Duality again.

TRIODES

The step from diode to triode semiconductor—the development of the transistor—proved to have tremendous consequences. Its results are some of the most interesting technological history written recently, including just about everything visible in applied electronics today and, evidently, there is much more coming. In 1948, nobody would have thought that the inefficient transistor exhibited by Brattain, Bardeen and Shockley could become the basis of many new applied electronics technologies. Yet, for all its importance, the transistor is fairly simple. It is basically a pair of semiconductor diodes, much like those described earlier, built back-to-back.

Since a diode has P and N sides, there are two methods of back-to-back assembly. These are termed "NPN" and "PNP"

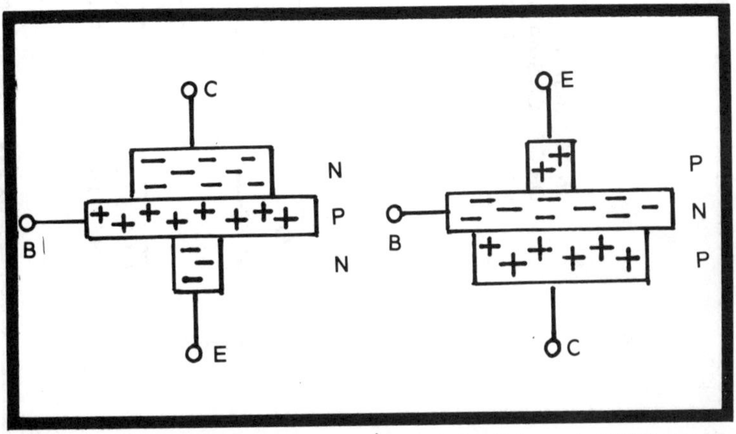

Fig. 3-28. General construction of NPN and PNP transistor triodes. These structures are always formed in single crystal germanium or silicon material, but there is a tremendous variety of detailed geometries.

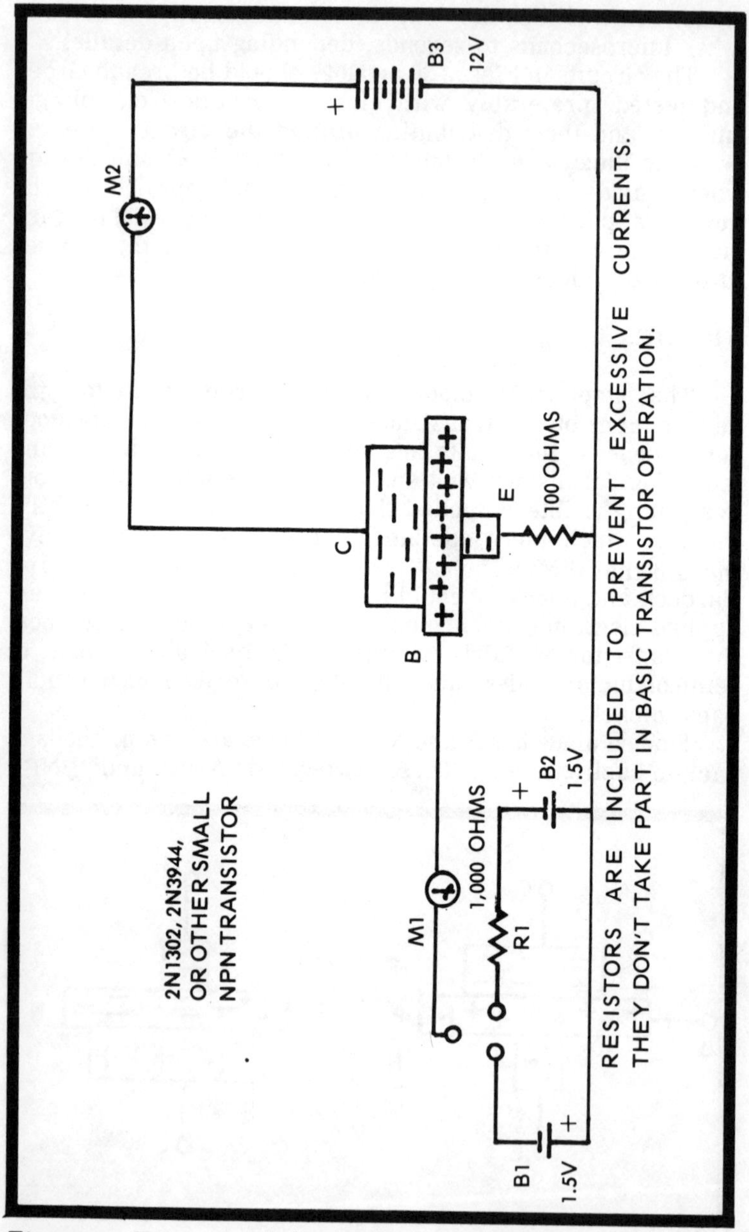

Fig. 3-29. This circuit illustrates the basic idea of how a triode transistor works. The base-collector diode is always reverse-biased, but it conducts if it can steal carriers from the forward-biased base-emitter diode.

as shown in Fig. 3-28. Each is a structure within a single piece of semiconductor crystal. The following discussion is based upon the NPN transistor, but with appropriate changes (converting to opposites, mostly) it applies equally well to the PNP transistor.

If an NPN transistor were sliced and magnified a few thousand times, it would look something like Fig. 3-28. This crystal block is divided into three regions: a small N-type region at the bottom, a large, very thin P-type region, and over this another N-type region appreciably larger than the bottom one. These parts are called simply "emitter," "base," and "collector." A PNP transistor appears here, too, with its dual nature being very apparent; both are drawn with charge passing from top to bottom as they will appear in schematics for normal application. The actual working parts of transistors are not so much the doped semiconductor regions as the junctions—the emitter-base NP junction and the nearby base-collector PN junction.

We are concerned now with the entire base-emitter circuit loop in Fig. 3-29. This consists of B1 or B2 and R1, the switch that chooses one or the other, meter M1 (to see what's happening) and the base-emitter PN diode. This diode can be either forward or reverse biased, depending upon whether S1 switches in a reverse bias (B1) or a forward bias (B2 and R1). This loop resembles the diode circuit described in the last section, but we notice that charge carriers moving in the base-emitter PN junction will circulate extremely near the base-collector PN junction. This proximity is of central importance.

Now we look at the emitter-collector circuit loop. Starting at the B3 positive terminal we proceed through M2 (to see what's happening) to the collector terminal, through the collector-base NP junction, the base-emitter PN junction, out the emitter, and return to B3. The collector-base PN junction cannot conduct any current, it seems, because B3 reverse biases this junction for either choice of 1.5-volt batteries. It looks as if no charge can move in this loop. If, while trying to forward bias the collector-base PN junction, we make the base 1.5 volts positive by switching in B1, then the base-collector diode sees 10.5 volts reverse bias and the base-emitter diode sees 1.5 volts reverse bias.

And it looks like if we switch in R1 and B2, the situation would be no better, although this would bias the base-emitter diode into conduction. Yet this might be a step in the right direction since it's already established that transistors work, somehow. If we actually do this, it turns out to be more than a small step. We observe a very large collector current. How does this happen? It happens because the reverse-biased collector-base junction is a very effective thief! It cannot conduct, using its own carriers, but if it can steal the correct polarity carriers from the base-emitter diode it conducts very well.

When we bias the base-emitter junction into conduction, holes flow from the base terminal and electrons from the emitter terminal to the BE PN junction. Earlier we talked as if the two sets of moving charges met exactly at the PN junction, but it's more true they meet in its vicinity—some electrons find their way into the P-type base and some holes into the N-type emitter. But in the NPN transistor, electrons that pass into the P-type base region are now very near the reverse-biased PN base-collector junction, and they are swept into it by the reverse-biased diode field. Notice that a reverse-biased condition for holes, which are normally in the base, means forward bias for electrons, which are supplied by the conducting base-emitter diode. In actual transistor operation, better than 95 percent of all the electrons flowing from the emitter toward the base NP junction are swept up by the collector field before they meet any holes in the base region.

This is the "mechanism" responsible for the way transistors work. It leads to two facts of transistor performance that will become important to us later. First, because most of the electrons get across the base into the collector region, it turns out that a small base current controls a large collector current—some ten to two hundred times greater than base current. And secondly, since electrons entering the base region are swept up by the collector, collector current depends not upon collector voltage but upon the supply provided by the base-emitter diode, provided collector voltage is great enough to carry out the sweeping process at all. For this reason a transistor collector terminal appears as a current source—a real-life Norton generator.

ELECTRIC LOGIC

Can electronic machines think? Some people insist this cannot possibly be, others point to examples such as computers that develop mathematical proofs. These machines work out consistent series of ideas, learning from experience, to develop logical, correct mathematical proofs. When men start with the same problems and develop the same results in the same manner, they are said to be thinking—and thinking well.

A new "adaptive control" technology features machines that are taught to do a job. As they work, they teach themselves to do the job better; these machines are found on production lines now. Some computers will listen to your words—and reply to you. Can machines think? It looks like the answer is "yes," and soon they will think better. Machines just don't develop personalities.

The tremendously complex electronic logic technology rests on a few simple basic sets of ideas. One of these is "digital logic," a system based on two extremes: on or off; high or low, zero or one. Such circuits must be combined in complex patterns to accomplish even basic computational work, and the development of large computer systems out of such basics ranks as a major human achievement. Yet digital logic circuits seem to have a certain weakness in working with only two values. Actually, this is a major strength, making such circuits preferable to any other way of doing the same jobs. It turns out that two-valued or "binary" circuits assemble into complex systems that use fewer parts than any other logic circuits. This surprising result has been reduced to a mathematical proof. And binary logic circuits minimize production costs, since they are composed of a very few basic elements, many times repeated. A complete computer control and arithmetic system can be built of only two kinds of circuits (NAND and NOR).

Also, binary logic circuits relate directly to a mathematical logic system known as "Boolean Algebra." (George Boole, 1847.) Effective binary logic circuits can be analyzed or developed for maximum effectiveness and computing capability without actually building the circuits. The problem of machine design becomes a problem in mathematics, and mathematics is an extremely powerful way to solve problems. The rapid development of modern computers couldn't have happened without this close connection into mathematics and the relative ease of transferring the mathematical engineers' results back into working circuits and systems.

And finally, binary circuits are more reliable than any other type circuits. Let us suppose we have a 12-volt supply line and we represent a binary one by +12 volts or something near it, and a binary zero by about 0 volts (positive logic). How much error can this circuit accept before it ceases to distinguish sharply between 0s and 1s in its own operation? That depends upon circuit design, but the margin for error is far greater than, for instance, in another circuit that works on 0, +3, +6, +9, and +12 volts. This latter arrangement looks more effective, but it not only takes more parts than a binary logic system, but much of the binary's voltage tolerance is

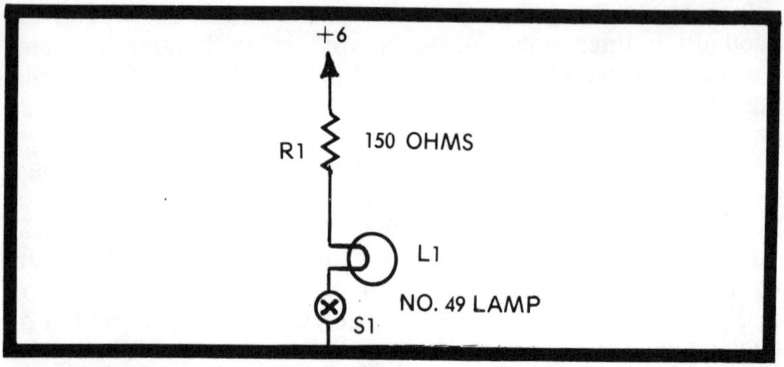

Fig. 3-30. Lamp L1 gives a binary indication of the on or off switch position. The switch here and in the following circuits could be a transistor.

now used up in several intermediate choices which must be hit fairly accurately.

These are some reasons binary logic circuits are used in all large computers (a few variations were tried in the early days of computer technology). Binary logic guarantees maximum utility, a remarkable, direct connection to the powers of mathematics and reliability. Such a mix would be a winner anywhere, and it went over big in computer technology.

We begin our examination with Fig. 3-30, the simplest possible binary circuit. This is simply an indicator. The binary (2-valued or 2-position) element is switch S1, whose positions are on or off. By adding a mechanical linkage, these positions might represent "window open," for instance, or "window shut." Lamp L1 shows us which setting we have. The no. 49 lamp takes 60 ma at 2 volts, and applying Kirchoff's loop theorem we see the 150-ohm resistor is a bit large. This is common practice in logic lights. We cannot see any difference between a lamp that is off and one that has failed (unless we take off the panel and make tests inside) so we run this lamp a little under specs. This improves life and reliability tremendously. If you get a chance to observe a real computer, notice the pale yellowish tint of its indicator lamps.

The circuit in Fig. 3-30 simply tells us if S1 is on or off. If S1 were situated somewhere distant from the indicator, this could be useful information. (Magnetic reed switches work well in circuits like these.) We can extend the utility of this circuit by adding another switch, which gives us the setup in Fig. 3-31. Here we get a lamp indication if either S2, or S3, or both, are closed. (How could you make that "but not both"?)

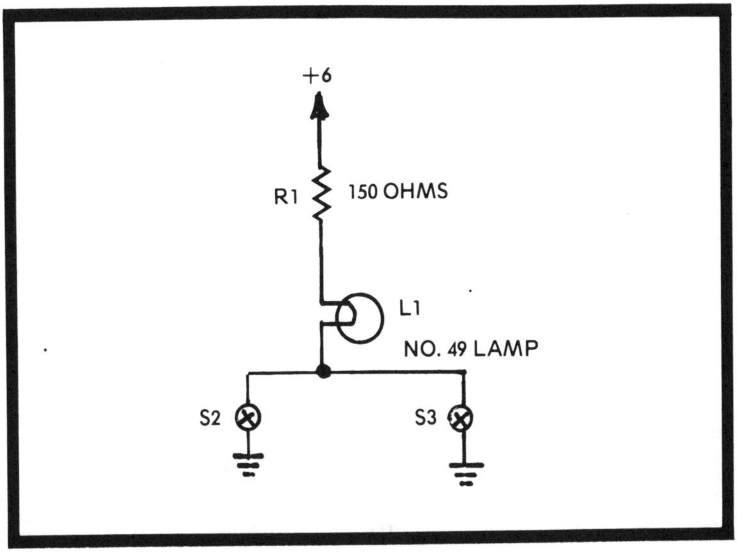

Fig. 3-31. Adding switches improves versatility. The switches don't have to be all at the same place.

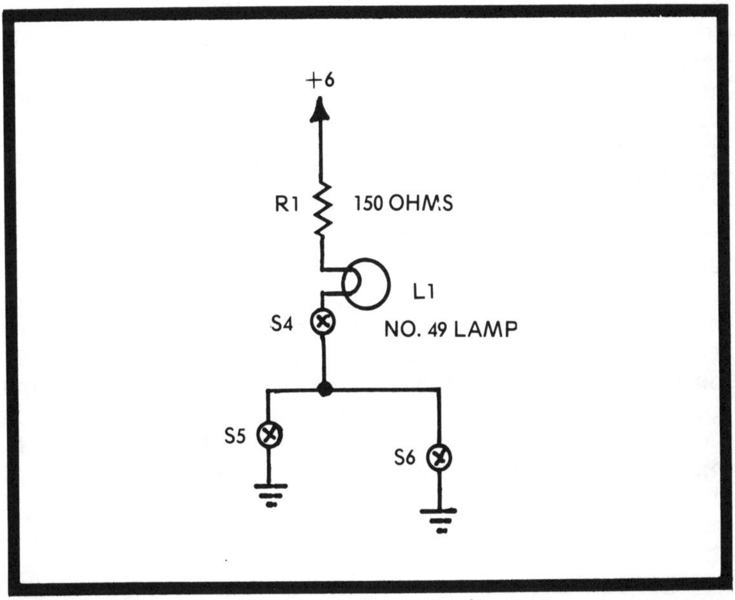

Fig. 3-32. Connecting switches in series and in parallel increases the number of conditions that will or won't light the lamp.

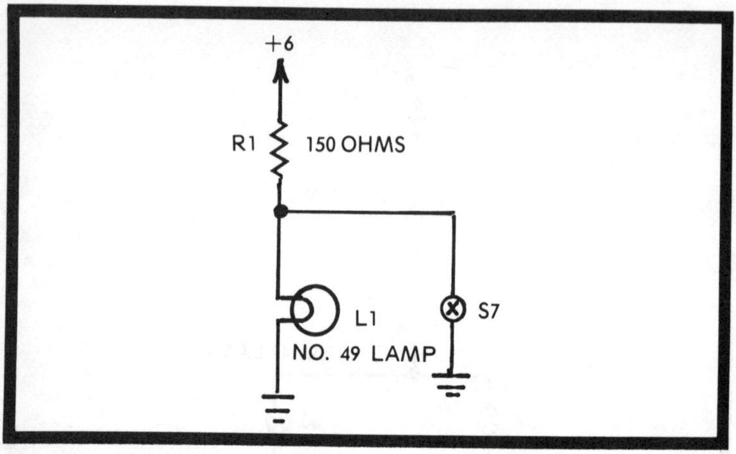

Fig. 3-33. Closing a switch can turn a lamp off as well as on. Contrary circuits like this are useful when designing circuits that work even if something has failed inside.

A more complex function appears in Fig. 3-32. Here, a lighted lamp signals us that S4 is closed, and so are either S5 or S6 (or both). Since we can install any of these switches to be either open or closed when a definite condition exists, this circuit is doubly versatile. For instance, if S4 in this circuit is mechanically linked with S1 (Fig. 3-30) so that one or the other is always on, then we get the following: If S4 is off, lamp L1 lights. If L1 is dark, then L3 comes on if S5 or S6 close. This is the beginning of a versatile system.

Finally, the circuit in Fig. 3-33 shows us how to turn off a lamp by turning on a switch. The 150-ohm resistor is largely responsible for the lamp circuit current. Closing S7 shorts the lamp but changes the voltage across R1 by only 2 volts. In computers and control systems the switches used here are replaced by electronic circuits whose transmission properties (on or off) depend upon operating signals, past events, and upon gating and control voltages. A computer-like example is the elementary but effective counter project in Chapter 9. In this system a number of logic circuits react to binary signal and control voltages to count the frequency or number of impulses of a variety of input signals.

SCHEMATICS

Up to this point, various diagrams representing circuits were to simple they spoke for themselves; besides, they were carefully labeled. Now we come to more complex diagrams. To begin with, what is the difference between a "schematic" and a "wiring diagram?"

Schematics vs Wiring Diagrams

Both schematics and wiring diagrams share a common purpose. Each shows how something is assembled. But from that point they diverge completely. A wiring diagram shows the parts in some electronic circuit and how they are connected together. The parts appear as little pictures of themselves—a transformer with visible core and even a strap around it; a wirewound resistor complete with mounting hole; a chassis or box drawn as if it were a real one with its sides swung partway down (the wiring diagram may have begun as a photograph made just this way). All this presentation supposes its user knows exactly nothing about how circuits and systems work; that he cannot comprehend even the simplest engineering description of the things he works with.

On the other hand, a schematic assumes its user does not have to copy a picture. It supposes he can absorb a **symbolic** circuit description, choose real parts without benefit of guidance from little pictures, and construct a functional duplicate of the original. Since no mechanical detail is presented, the schematic is very much simpler than a wiring diagram, and this opens the way to communicating real engineering facts.

Purpose Of Schematics

A good schematic is the simplest possible drawing that communicates to the reader (1) how the circuit works; (2) the time sequence, if any, of interval events; (3) the functional organization of the system; (4) the values and ratings of the individual components; (5) the interconnections between components; (6) key facts of circuit adjustment and performance.

Meeting these six general requirements is far easier than it might seem, since the purpose is communication, and whoever aims for this goal stands a fair chance of making it. It often happens a man drawing a schematic is really talking to himself, and he develops something that doesn't readily carry facts into another man's mind. Examples abound. The worker who pulls himself out of this and aims for communication gets at least a valuable side benefit. His effort in clear thinking improves his work—the effective communication becomes almost a side benefit, yet a reliable one.

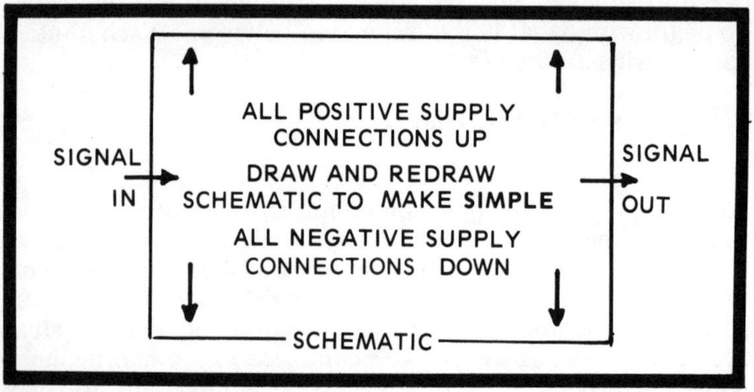

Fig. 3-34. Establishing an overall general pattern before getting down to details will improve the finished schematic's read-ability.

Who Needs Schematics

Students need schematics, since each drawing (if well done) is a concise summary of all important facts about a working electronic circuit or system. This puts all the designer's work at the student's hand; then study and practice lead to real comprehension and, in time, mastery of design skills. The wireman needs schematics since they provide him with all facts needed to build the system. (Mechanical details are commonly communicated separately, if they are important.) If he cannot build a working model because the schematic is incomplete or unclear, valuable time is lost.

The serviceman needs schematics since they provide him with overall pictures of how circuits or systems work, what their interconnections are, what component loads and working voltages appear, and so on. This enables him to observe a detail of system performance and work back to the circuits whose function or malfunction are responsible for it. If he makes adjustments, a look at the schematic enables him to see what his changes are achieving inside the system. Finally, the professional needs schematics because they help him think and to communicate his results to other professionals. His picture of what he seeks to achieve closely resembles a schematic.

Schematic Conventions

This book's schematics hardly illustrate the communication of complex ideas. They are too simple. We are concerned with fundamentals; yet the principles are the same in any case. First, we want our schematics as much alike as possible in order to spare the eye unnecessary, repetitive details while emphasizing key facts. We work according to an overall plan of charge flowing from top to bottom (gravitational schematic) and signals or sequence of events passing from right to left. See Fig. 3-34. About half the circuits come out apparently inverted, since we work with PNP (collector goes -) and NPN (collector goes +) transistors. This convention is rare, outside some industrial and scientific fields, but proves very valuable in visualizing the action of many signal, industrial and pulse circuits.

We avoid long lines around the schematic. If two points are connected though well separated on the drawing, a few words or symbols say so; long lines add nothing to the facts communicated by the schematic. Similarly, transformer windings on the same core are related by their labels. A complete transformer symbol appears only where it fits in easily with the rest of the drawings. Lots of circuit hints are left around for the convenience of the reader who wants to really analyze the circuit. Computed values are labeled (est) until observed. Enough voltages and currents, shapes of pulses, and relevant notes should be included so that the user can work out essential details without resorting to very much higher math. Omitting this centrally important, although somewhat supplementary material, is a kind of shoddiness, and it may tend to be worse than that if the schematic writer hasn't proved his circuit. Finally, we use no draftsman's artifices. If two lines are connected, they do not cross; they make a T. And if they cross they aren't connected. Dots can get lost, and the most careful checker goofs occasionally. Once a man who understands a circuit uses these conventions to draw it right, error is unlikely to turn up later. See Fig. 3-35.

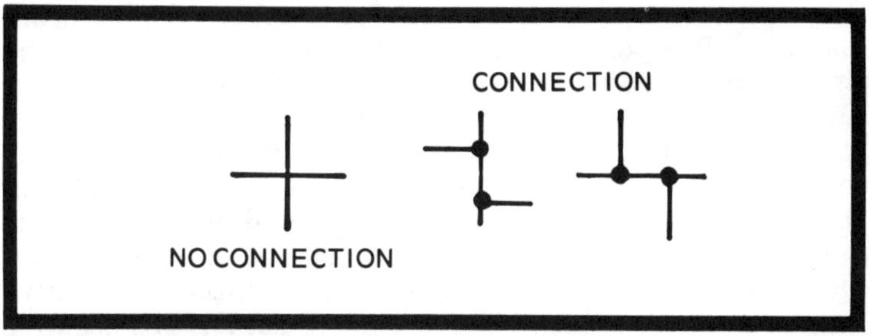

Fig. 3-35. Simplest conventions are best, and the man reading or copying intersections like these isn't likely to goof.

Resistor values are chosen at 10 percent "magic number" values, and are half-watt 10 percent tolerance types unless otherwise indicated (modern industrial schematics may fix on quarter-watt 5 percent units). Composition resistors are assumed; it's up to the reader to decide if he can or cannot use wirewound resistors. A WW says the designer **wants** wirewound; MF denotes metal-film, perhaps for their tiny inductance or for low-noise application. If a resistor is rated 5 watts or more, it must be wirewound; composition types do not come over 2 watts.

Inductors appear rarely, compared to resistors and câpacitors. Their values are given much like capacitor values, but additional data commonly appears to supplement the unit's electrical value.

Capacitors are good quality disc ceramic or tubular unless otherwise clearly indicated ("MICA" means mica, etc.). Electrolytic capacitors are indicated by a "+" at one terminal and a working voltage specification. Capacitor values appear in picofards, nanofarads, or microfarads. For details on component value notation see the Appendix.

Schematics And Circuits

Fig. 3-36 is the schematic of a simple utility amplifier. It was breadboarded with 2N1304 and 2N1305 germanium transistors, but isn't sensitive about transistor specifications. This schematic presents all facts needed to understand the circuit, or to build it. It uses words as well as symbols, each where best suited. Here, we are particularly concerned with how the schematic relates to the circuit, and we follow it as if we were actually building the circuit. If you actually do that, this circuit goes well on a piece of perforated board or strung up in the air breadboard fashion. When building on a board, the parts are all laid out with an eye to some sensible order that minimizes parts interconnection lead lengths; if up in the air the parts are simply added one by one. Since this is a very simple circuit, its schematic serves as a parts list.

We wire this circuit by making the parts interconnections shown in the schematic. First, we locate all the ground connections and wire these together. That will be an input terminal, one end of D1, the Q1 emitter, an output terminal, and the battery return lead. Then we interconnect the positive supply terminal, one end of R3 and one end of R4. The negative supply terminal goes to one end of R1 to the Q2 collector.

Now the audio input terminal is connected to C1 and nothing else. The other end of C1 is connected to the Q1 base, with an end of R2, the remaining end of R1 and the banded end of D1. The free end of R2 is connected to the Q1 collector, the Q2 base, and the free end of R3. The free end of R4 is interconnected with the Q2 emitter and the positive end of C2. The remaining end of C2 is connected to the audio output terminal.

In this work we never supposed any line in Fig. 3-36 stood for a wire. The lines merely showed which ends of which components were connected together or to terminals. We used this information to put in wiring as convenient (experience and a real understanding of how circuits work are tremendously valuable) and when no interconnections remain to be done the job is finished. On a large job it may be helpful to put a tiny check mark by each interconnection as it is finished; this saves scribbling all over the schematic.

The final step in construction is proving the circuit. A test under power shows voltage levels as indicated, and if these aren't right it's hardly any use doing signal tests. With everything in order, a small input signal with a meter or scope output proves whether or not the circuit is managing signals properly. Usually, input and output detail characteristics aren't important. The construction job is done.

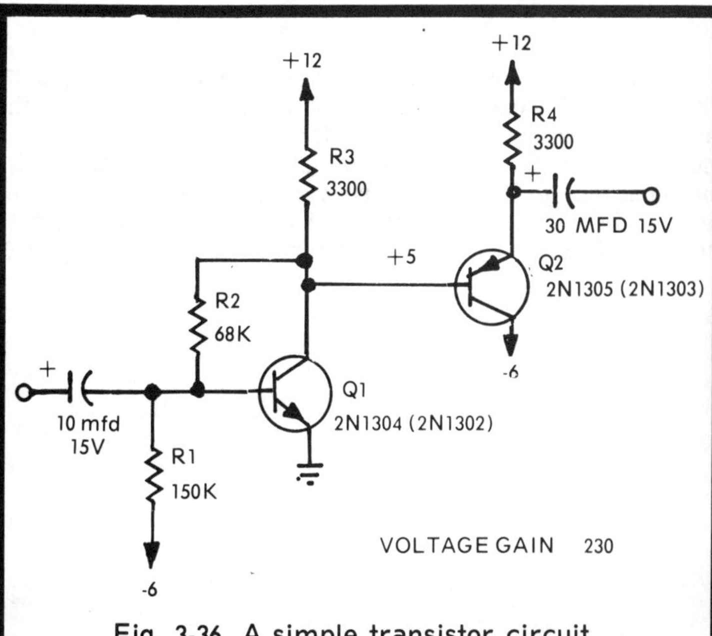

Fig. 3-36. A simple transistor circuit.

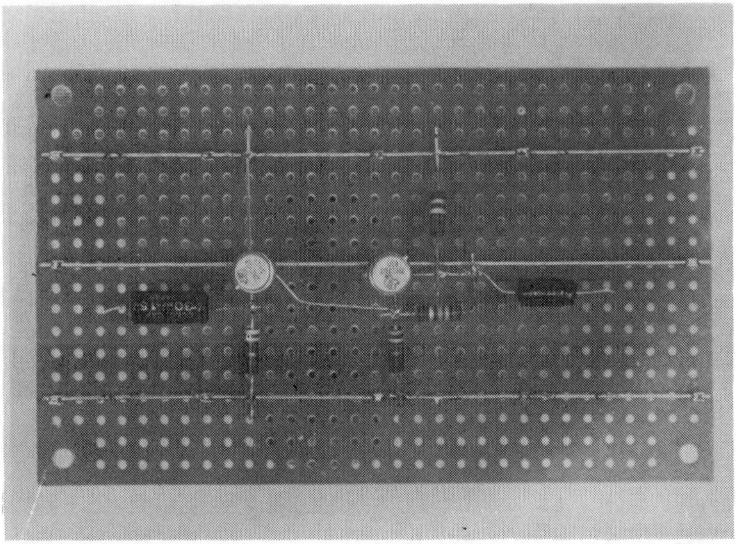

Fig. 3-37. Modern electronics circuits often look very much like the schematic. This makes things nicer for the designer, the builder, and the user.

A photograph of the finished circuit appears in Fig. 3-37. This is a semi-breadboard assembly, nice to have around the lab a few days, but not meant for long-term work. If you trace the interconnections you will find it corresponds exactly with the schematic in Fig. 3-36. Here, we were concerned with the relation between a schematic and a real hardware circuit. How the circuit works comes up for discussion in Chapter 6.

Chapter 4

Diode Circuits

If they laughed when you sat down to a bench full of diodes, they knew little about modern electronics. Diodes are very versatile. The good circuit engineer finds many uses for them beyond rectifying AC into DC. In fact, a quick test of an engineer's skill and ingenuity is to look at his work and see how he uses diodes. From a circuit viewpoint, the diode's leads connect right into the diode's fairly complex physics; the designer who uses his physics knowledge creatively shows he has a good understanding of what really goes on in circuits.

Diode physics is an important subject. You will learn to constantly look out for new facts about diode physics, and what you learn will work for you. Transistors, for example, are basically diodes. So are "4-layer" devices such as Shockley diodes and SCRs. Field-effect transistors and unijunction transistors, tunnel diodes, the recently developed laser diodes, all are basically diodes. Diode theory is as important to modern electronics as a building's foundation is to the building: if it's weak, you're in trouble.

Diodes find lots of work protecting circuits. The moment a circuit is first turned on and before voltages settle to their operating values, is often a hazardous instant from the circuit's viewpoint. Another occurs when the circuit is turned off. Overloads sometimes destroy circuits; so do overlarge input signals and forseeable but not really predictable events inside the circuit. Well-placed diodes help the circuit over these rough moments, which improves reliability tremendously. Diodes do many other good things. They can be assembled into logic circuits that perform complicated switching operations. You can use them for excellent voltage regulators, and to tune circuits. There's no doubt about it: diodes are important. Large as it is, this chapter is pretty small for its subject.

A GENERAL LOOK AT DIODES

The collection of things we call "diodes" breaks down into silicon and germanium, signal and power, point-contact and

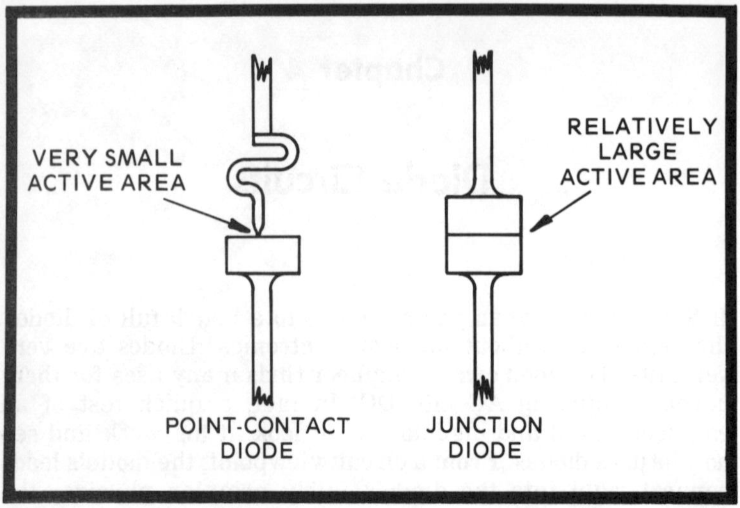

Fig. 4-1. Point-contact and junction diodes are basically alike, but the junction types can carry larger currents (and typically work slower).

junction types. We look at these general and very basic differences first. Then we go on to diodes in forward conduction, diodes in turnoff, and the temperature dependence of diode characteristics.

Diode Materials

Diode PN junctions are formed in tiny pieces of either germanium or silicon crystal. A few modern extremes go into gallium arsenide and other exotic materials, but these won't concern us here. The germanium diodes operate on a smaller voltage scale (0.3 volts) than the silicon types (0.7 volts) but tend to leak more than the silicons (microamps vs nanoamps for silicons). The silicon diodes, since they are less leaky than germaniums, work at higher temperatures, but some germaniums switch faster than most silicons. In most room temperature applications, there is no compelling reason to prefer one over the other. Look in your parts box to see what you have and then design your circuit to use it.

Diode Types

Diodes fall into two classes: signal diodes and power diodes. Signal diodes work at very small voltages and currents; power diodes manage large forward currents and, often, high reverse voltages. Their active structures, or PN junctions, reflect this difference in intent: a power diode is a rather large structure, but it may take a microscope to locate the working area of a signal diode. Signal and power types overlap in a third class of interest to digital circuit workers—switching diodes. Yet whatever its purpose, a diode is basically a PN junction. From this you might guess a small power diode could find work in a signal circuit, or a rugged signal diode might make a nice power rectifier if not asked to do too much. That's true.

Still another difference between diodes is whether they are point-contact or junction types. (See Fig. 4-1.) This is a manufacturing difference that once was important to circuit designers. These days junction diodes can be made so small they act like point-contact types. The details largely interest surplus semiconductor users.

Forward Conduction

A register's voltage-current curve comes out a straight line, as shown in Fig. 4-2. Not so the diode, where the voltage dependence on current changes as current increases (we are thinking in Norton equivalent terms). Another way of saying

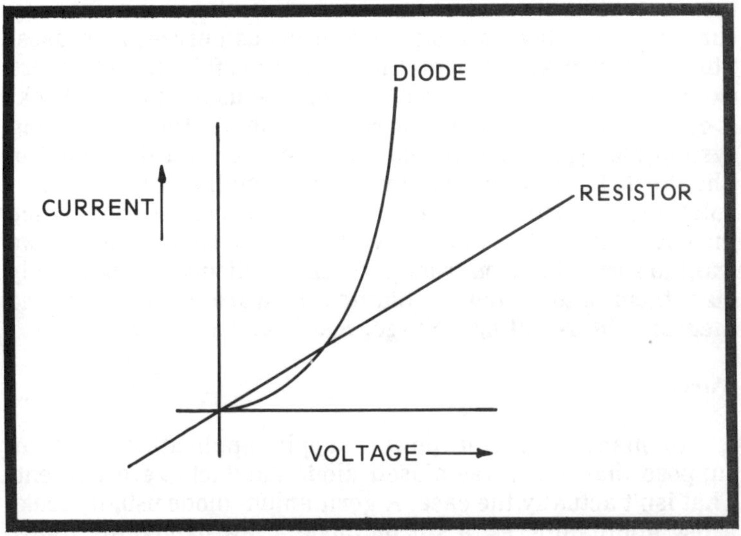

Fig. 4-2. The current through a resistor depends upon voltage, according to Ohm's Law. Diode characteristics cannot be represented by such simple straight lines.

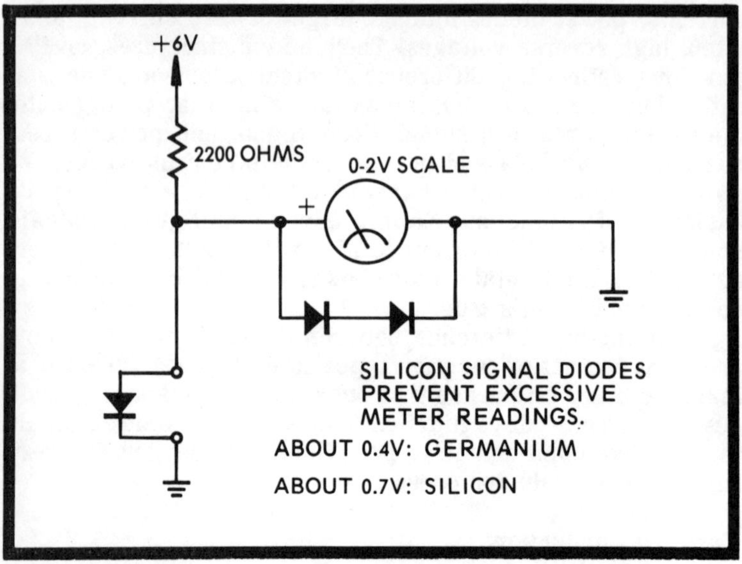

Fig. 4-3. This simple circuit provides a way to sort germanium from silicon diodes and to spot which way they conduct.

this is that the resistor's resistance is the same for any current, but a diode's resistance drops as current increases. This characteristic is sometimes very useful (and at others extremely inconvenient), but we won't be using it in this book. It could be used to make up a remote temperature measuring system along the lines of the thermistor circuit described in Chapter 1. If we are working with very small signal or control voltages the germanium diodes give us a design edge, since they begin to conduct at lower voltages than silicons. And if we want to sort out silicon from germanium diodes, we need only bias them into a few milliamps forward conduction and measure the resulting voltage. See Fig. 4-3.

Turnoff

In many signal or power supply applications we can suppose that a reverse-biased diode conducts zero current. That isn't actually the case. A germanium diode usually leaks a few microamperes, a silicon diode a few nanoamps; these leakages increase with applied voltage (and temperature). See Fig. 4-4. Here, we see that germanium diode leakage increases gradually with increasing reverse voltage. And we notice a silicon diode's leakage remains very small until some critical reverse voltage is reached, then it zooms to very large values. A diode in this operating region is said to be in "breakdown." Not all diodes in breakdown are very happy about it (they may develop localized hot spots in their PN junctions, leading to early failure), but one class of diodes is designed to work in breakdown. These are the zener diodes discussed later.

Diode Temperature Dependence

The curves of Fig. 4-4 show that a definite typical voltage exists across a diode when charge is flowing through it. This voltage tends near 0.3 volts for a germanium diode and 0.7 volts for a silicon diode, and as temperature rises (an electronic chassis warming up, for instance) this voltage falls.

It turns out that both germanium and silicon diode voltages drop about 2.5 millivolts per degree C temperature rise. Since a realistic estimate of room temperature operation is 20 degrees C, nominally plus-minus 10 degrees C, we estimate that our apparently fixed diode working voltage varies about 25 millivolts each way. This doesn't include natural differences between individual diodes of the same

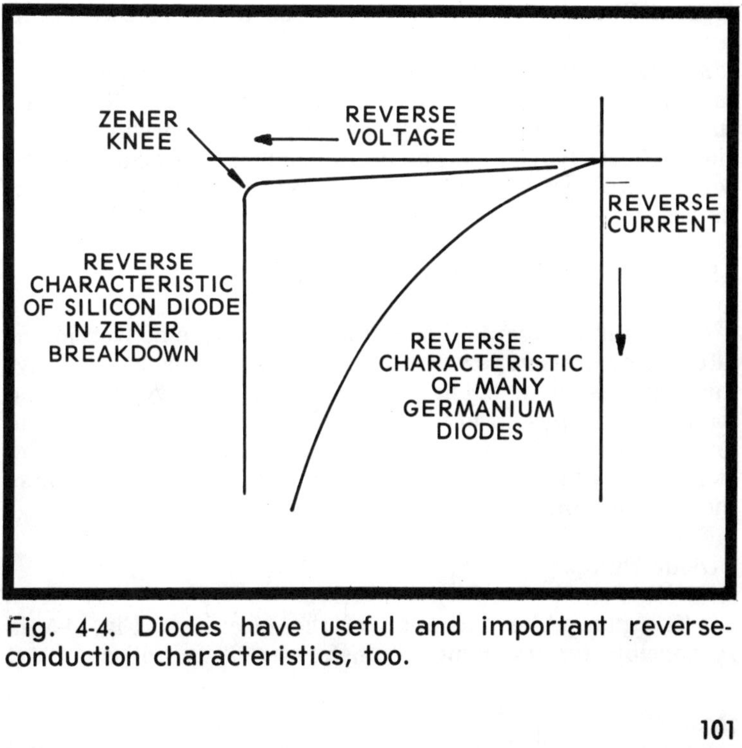

Fig. 4-4. Diodes have useful and important reverse-conduction characteristics, too.

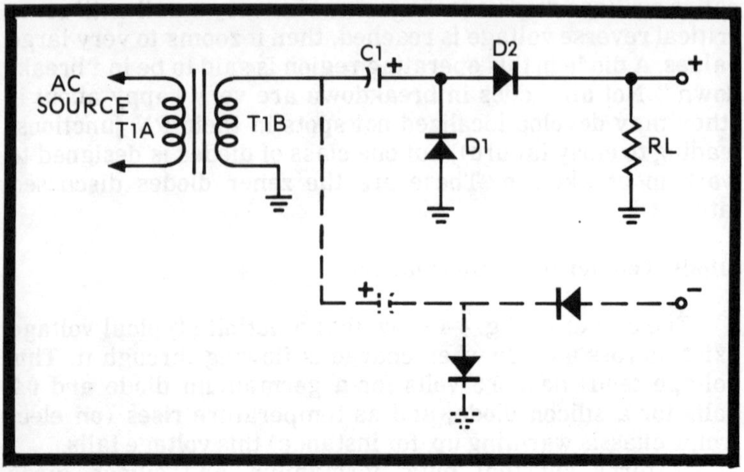

Fig. 4-5. This "diode pump" circuit works remarkably like an old-fashioned water pump. Two diode pumps can supply opposite polarities from one transformer.

type. And for some applications, working temperature would range from winter cold to summer auto-body hot (-30 to +70 degrees C) so that the 2.5 millivolts per degree rule of thumb needs correcting. Accordingly, the least temperature drift we can expect is maybe 10 percent in germanium and 5 percent in silicon diodes. Our diode circuits won't be upset by this drift, but transistor base-emitter junctions are also forward-biased diodes. This drift and what to do about it is a centrally important problem we will face in Chapter 6 on biasing transistors.

THE PUMP AND BRIDGE RECTIFIERS

A common, simple diode application is the conversion of alternating current power into direct current power. Usually, the voltages involved are much larger than typical diode forward voltages, and currents far exceed typical diode leakage currents. As rectifiers, diodes look like very good switches. We begin by examining two useful rectifier circuits: the "diode pump" and the "wheatstone bridge."

"Diode Pump"

The pump-like operation of the circuit of Fig. 4-5 is responsible for its name. Transformer T1 supplies an AC voltage, and we can view its operation as sweeping the charge one way and then the other through its secondary, T1B. If the first half cycle sweeps the charge down through its secondary, this charge flows to ground, returns through D1, fills C1, and returns to T1B. No current appears in RL.

In the next half cycle, the charge sweeps the other way through T1B. It flows from ground up through T1B, through C1, forward through D2 and into RL. Since C1 was charged in the previous half cycle, RL sees a voltage equal to the sum of those across C1 and T1B. Since C1 was charged to the T1B peak voltage in the first half cycle and the T1B peak voltage is added to this during the second half cycle, in power supply work the diode pump appears as a voltage doubler. Notice since T1B and RL carry the same power but double the voltage operation, T1B provides double the current in RL. T1 will run very warm if you miss this elementary point in a power supply project.

Wheatstone Bridge

A very useful instrument circuit also carries this name, but it has no other relation to the following, except it, too, can

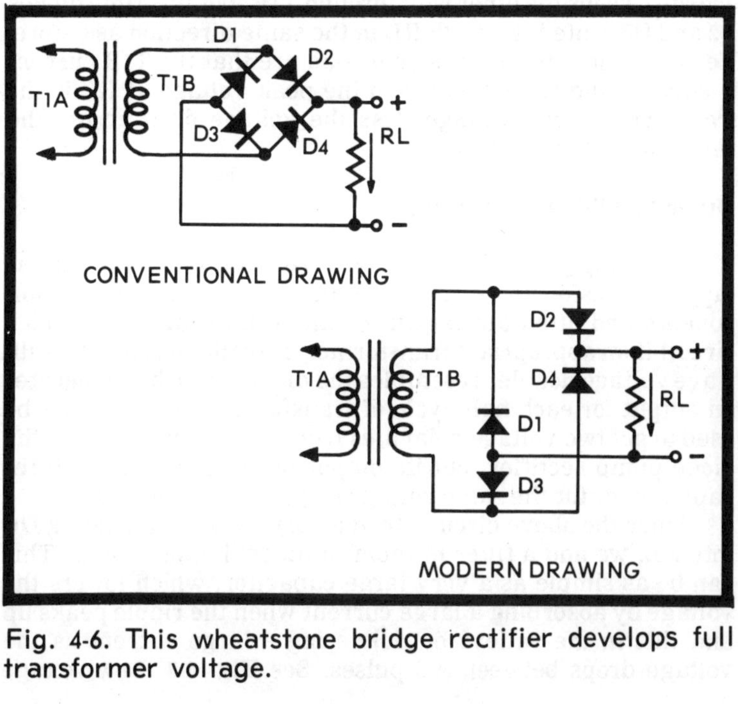

Fig. 4-6. This wheatstone bridge rectifier develops full transformer voltage.

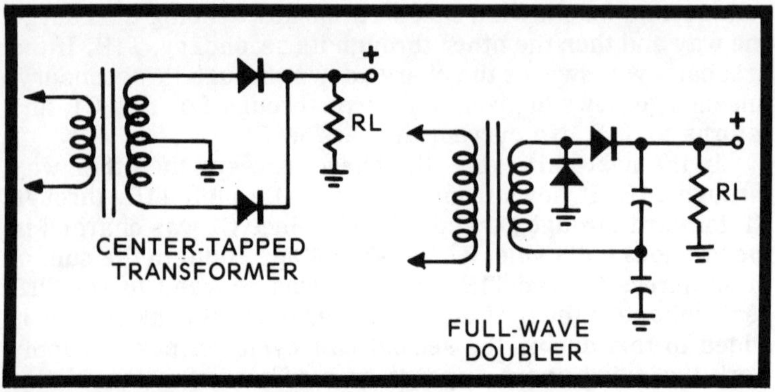

Fig. 4-7. Two other popular transformer and rectifier circuits. The full-wave doubler develops double the ripple frequency of the diode pump, so that only one half of the filtering is needed.

be drawn as a square resting on one corner. See Fig. 4-6. Here, we suppose the charge sweeps down through T1B, and we trace the flow through D4, RL (in direction indicated by the arrow,) and returns to T1B through D1. In the next half cycle charge sweeps the other way through T1B, but this time diodes D2 and D3 route it through RL in the same direction as before. Key points in this circuit's operation are that the T1B current is equal to the RL current, and the peak voltage at RL is the transformer peak voltage, less the voltage drop across the resistance of the circuit.

Other "Full-Wave" Circuits

The supply circuits in Fig. 4-7 are often seen. The center-tapped transformer circuit minimizes the number of components and often appears in commercial gear. It is a nice circuit if an appropriate transformer is available. And the full-wave voltage doubler resembles the diode pump but generates an output for each half cycle of transformer AC. It cannot be used to get two voltage polarities from one winding, as can the diode pump rectifier, but its output needs one half the filter capacitance for the same voltage ripple as a diode pump.

Since the above circuits feed a very rough or pulsating DC into RL, we add a filter to them in all real applications. This can be as simple as a very large capacitor, which lowers the voltage by absorbing a large current when the ripple peaks up and fills in the valleys by delivering a large current as the voltage drops between AC pulses. See Fig. 4-8. Low-voltage solid-state supplies usually improve the voltage appearing at the capacitor terminal by adding special regulating circuits between the filter and RL.

Power Supply Design Suggestions

A transformer, diodes, and filter capacitor picked up and wired as shown in Fig. 4-8 might work well, but probably not. The good designer concerns himself with details—all of them. First, what voltage rating is appropriate for the capacitors? We should choose something in this world. An over-stressed capacitor cannot be trusted; an excessive rating isn't economical. The solution is to work out a minimum rating and

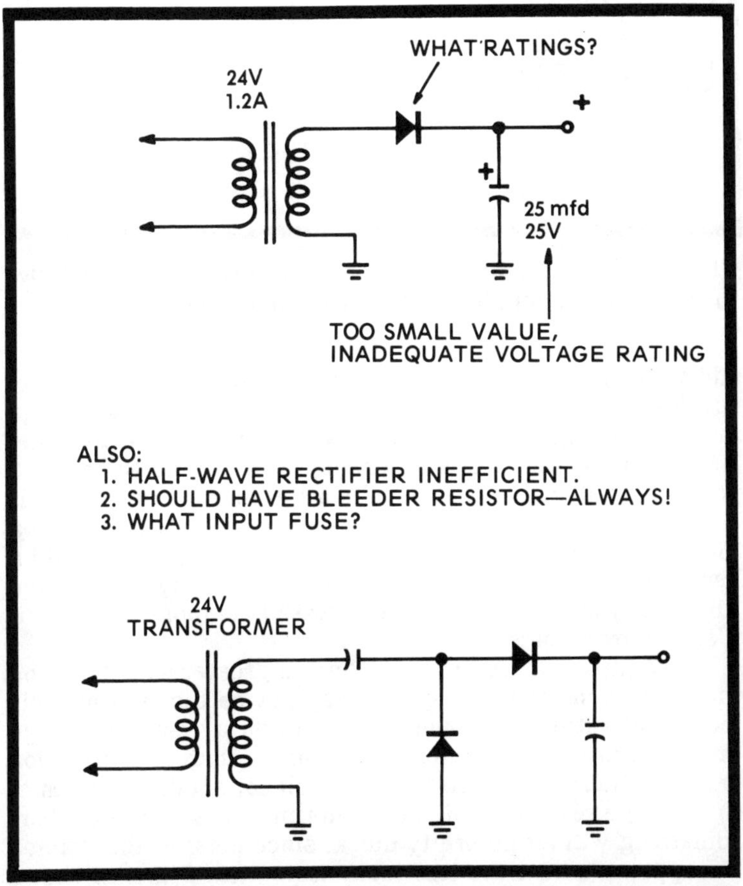

Fig. 4-8. A power supply circuit design that is started right but needs more work.

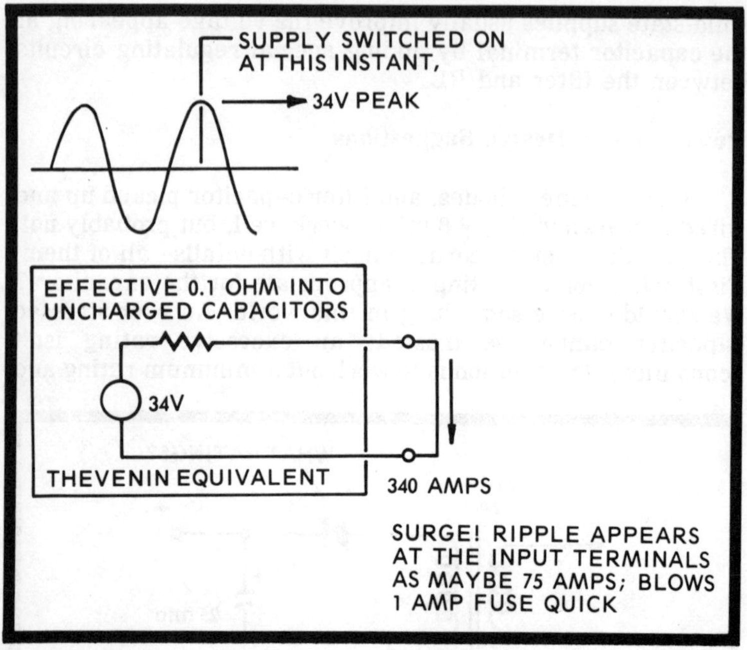

Fig. 4-9. This circuit illustrates how power supplies blow fuses though operated within their ratings.

add a safety factor of perhaps 25 percent to that. For instance, our diode pump circuit can deliver the full T1B peak voltage into C1. See Fig. 4-9. Peak value of a sine wave is 1.41 times the RMS value, so that a 12-volt winding could fill C1 to about 17 volts; this will go to about 19 volts if the supply line voltage is above specifications. A 20 percent safety factor brings us to a 25-volt operating rating; not bad for a 12-volt circuit. Would a lower voltage rating get by? Usually, yes. It might last some time, but you cannot trust overstressed components, even if it's your own doing.

Two capacitors in series will have a greater voltage rating than either one of them—but it may not be the sum of their individual ratings! It's necessary to add a resistor voltage divider across the capacitors to guarantee correct distribution of the applied voltage. Given two 12-volt capacitors in series, if the leakage resistance of one is ten times less than the other, something will let go pretty quick, since most of the applied DC appears across the less leaky capacitor. Fig. 4-10 shows the remedy to this problem. This equalization is necessary in diode circuits too, since reverse voltage will be distributed along a string with the leakiest diode seeing the least of it. Resistors applied this way in high-voltage circuits should not see more than 300 volts or so per resistor.

Now, let us look at something that may happen the instant a power supply circuit is turned on. At this instant, the filter capacitor contains zero charge; the power supply sees it as nearly a short circuit until it is filled to somewhere near operating voltage. A quite small, low-power supply may deliver several cycles of near 100-ampere current into its capacitor, leading to a tendency to blow fuses. This isn't very good for the diodes either, if they're rated conservatively for normal operation.

Three remedies present themselves, as indicated in Fig. 4-11. The most economical is simply a smaller filter capacitor (an active regulator or a zener diode improves this DC to electronic standards anyway, usually). A thermistor, whose resistance drops as it becomes warm in operation, is an excellent arrangement. Finally, a current-limiting resistor in the primary or secondary, shorted by a small relay a few seconds after the supply is turned on, is an effective, no-nonsense provision. Supplies up to the 1 to 10 ampere range (purposely vague, that spec) can get by without this detail attention; larger ones probably better with it.

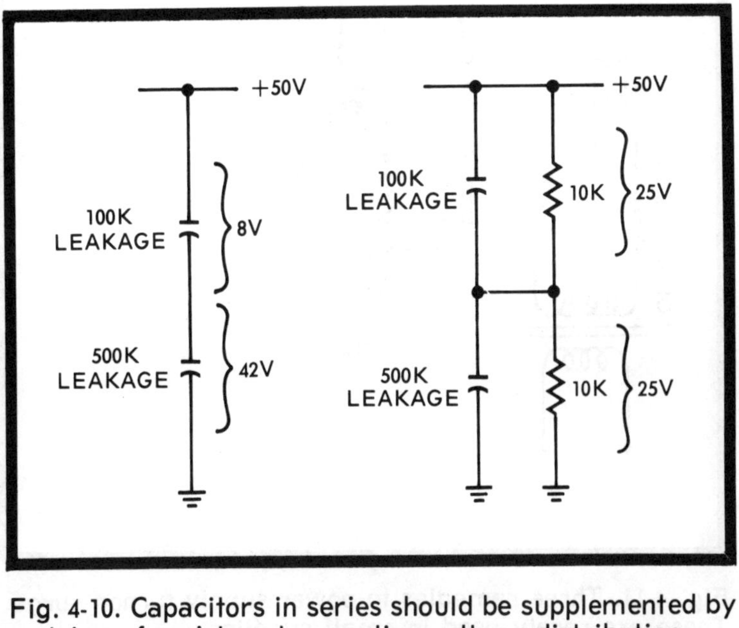

Fig. 4-10. Capacitors in series should be supplemented by a string of resistors to equalize voltage distribution.

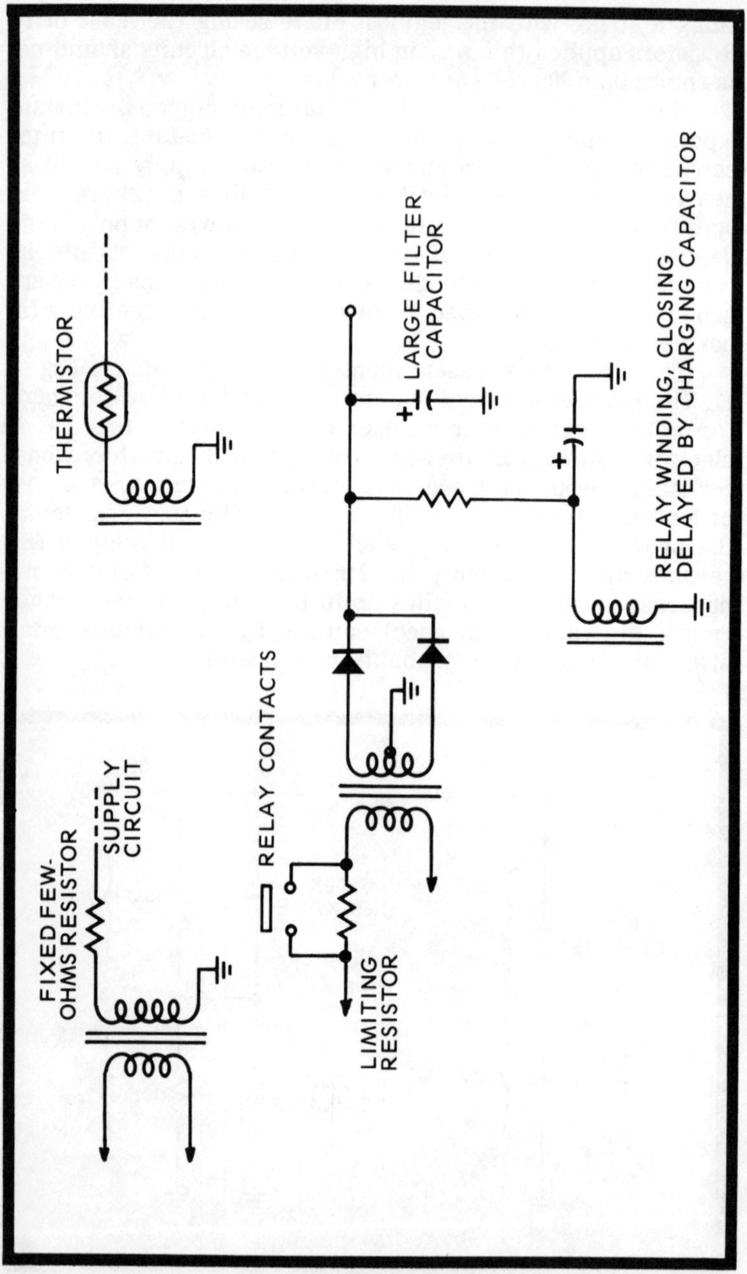

Fig. 4-11. Three remedies to power supply turnon surge. These are rarely used in small supplies.

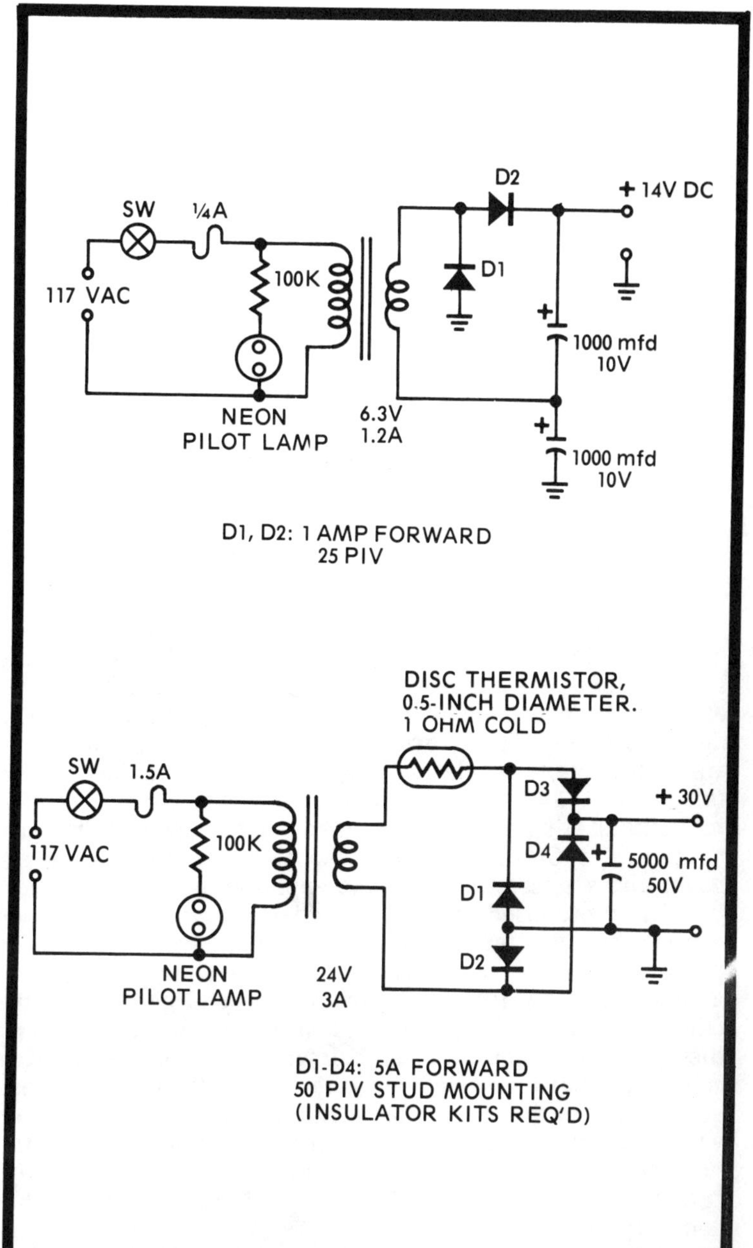

Fig. 4-12. Two correctly designed utility power supplies. Notice the fuses, always used with AC equipment.

In Fig. 4-12 there are two working circuits that apply the above basics. The first is a low-current diode pump circuit using inexpensive components. The second delivers more power; it uses a thermistor to control turn-on surge. Two other details also deserve comment. Both supplies are fused conservatively. The fuses are rated about 30 percent above primary current when the supplies are delivering rated output. These fuses are rarely needed, but when accidents occur—as they do in the most carefully run lab—it turns out that fuses are far cheaper than circuits.

Neither of these supplies has a sizable filter resistance in its output circuit. Some do and this resistance turns up full value in a Thevenin equivalent as part of the output resistance. In other words, it spoils regulation and it may upset the circuit powered by the supply. If a large capacitor cannot adequately reduce ripple, perhaps a zener regulator can do the job.

DIODES AS REGULATORS

Even a moderate reverse voltage produces a strong electric field in a diode PN junction. What happens if the reverse voltage is increased to a point where something gives and the diode conducts? Does this destory the diode? The diode survives if not overheated, and thereby shows a very valuable voltage-regulating property. Diodes engineered for this work (and others applied without benefit of the manufacturer's blessing) are called "zener diodes." They are named after Clarence Zener, who observed reverse breakdown and suggested a plausible explanation in 1934.

Zener proposed that PN diodes in reverse breakdown conduct because the strong electric field (over 250,000 volts per inch) literally tears electrons away from their parent atoms, and the free electrons carry charge across the otherwise empty PN junction. This is the "field emission" theory, and it turned out about half right. In 1954, K. G. McKay suggested that free electrons (there are always a few) are accelerated in the strong electric field, and when they collide with fixed atoms they knock off more electrons. This is the "collision ionization" theory, also about half right. Actually, both explanations (physicists say "mechanisms") apply, but under different conditions. Not all PN junctions go into breakdown at the same voltage. Zener's mechanism is at work when breakdown occurs beyond 6 volts; these zener diodes show a positive temperature coefficient; that is, their zener voltage rises as temperature goes up. And McKay's mechanism applies to diodes in breakdown at less than 6 volts, giving them a negative temperature coefficient. For a zener that regulates near 6 volts both mechanisms apply; their temperature drifts cancel, and the diode regulating voltage is stable against temperature change. No well-designed drift cancelling circuit is as effective as using 6-volt zeners with little drift to cancel.

Reverse biasing a diode into breakdown doesn't harm the diode if the voltage is within reason: the manufacturer's dissipation ratings. The high field strengths and strongly accelerated electrons are not enough in themselves to upset the PN junction's crystalline structure—excessive temperature does that. Some diodes have little imperfections in their structure, which leads to hot-spots in zener breakdown, but others make nice zeners whether their maker intended that or not. General Electric even recommends some of their transistors as low-priced zeners, with the base-collector diode going into breakdown near 9 volts. A small diode or transistor may dissipate 100 milliwatts; a large industrial unit can manage 50 or 100 watts.

The curves in Fig. 4-13 illustrate how zeners serve as regulators; Fig. 4-13 is Fig. 4-4 turned around and drawn in more detail. This curve was taken from a surplus zener by feeding various currents into it and observing the resulting

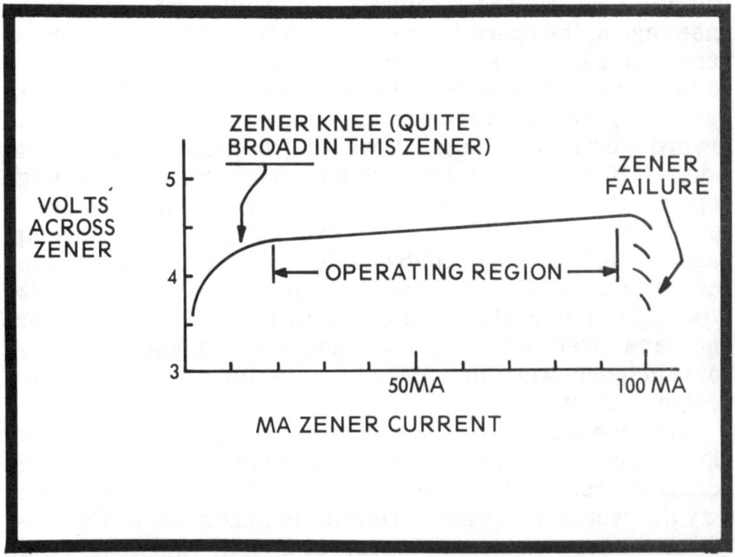

Fig. 4-13. Operating characteristics of an inexpensive zener tested to destruction. A spoiled zener is useless for any purpose.

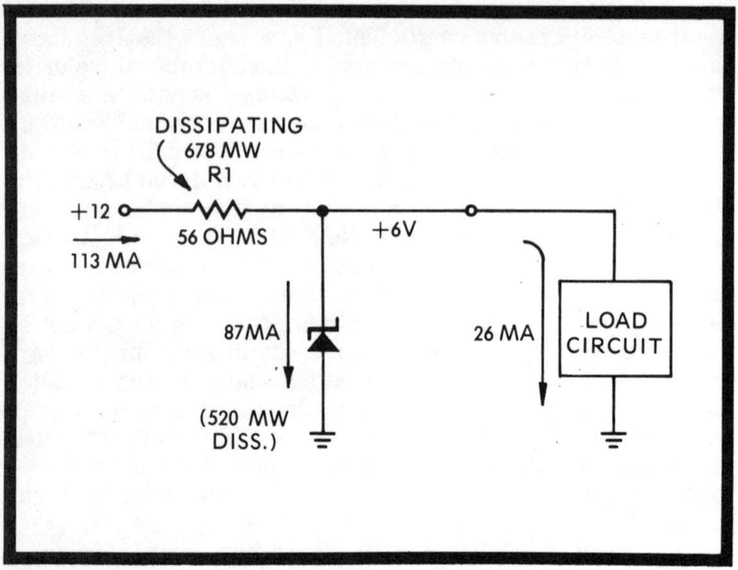

Fig. 4-14. A simple zener circuit at work. Current through the 56-ohm limiting resistor is almost constant, but it is shared variably between load and zener.

voltage. Four regions of interest are insufficient current, the knee region, the operating region, and the overload region. At very small currents the feeble breakdown process yields a voltage that depends heavily on zener current. As current increases, the physical processes become stronger and tend toward a definite voltage across the PN junction and its outside terminals. This is the "knee region," where the voltage dependence on current weakens. The manufacturer's specifications usually avoid a knee figure, since this apparently is not very controllable. Instead, a lower-value current that is above the knee for all devices of a particular type is given. In real life it's well to pick up a few extra zeners and save those with very low-current knees, since they will come in handy when that extra bit of performance will solve a design problem.

Now we come to the operating region. The well-energized diode now shows good regulating properties—the voltage across it is relatively unaffected by the current through it. The way this works in a practical circuit is illustrated in Fig. 4-14, where we see the incoming current is split two ways. Some goes to the diode and the rest into the load. Since the zener diode fixes the voltage at its end of the resistor and the supply determines the voltage at the resistor's input end, the current through R1 is fixed. This means the regulating circuit works by letting the load steal current at a constant voltage from the zener, and that regulation fails when the load starves the zener into its knee region.

Finally, we come to the overload region. Here, the zener's dissipation capability is exceeded; the PN junction overheats and doping atoms begin to move around. The zener's regulating voltage now drops, its regulating properties are degraded, and it soon fails completely. This condition is irreversible; the zener now has no value.

Zener regulator design is basically simple. The zener must get enough current to stay out of its knee region, but not so much that it ever goes into overload. Complicating factors are typically incomplete zener specifications, a plus-minus 10 percent or more line voltage variation and, probably, some irregular change in load. These matters are usually managed by a look at worst-case conditions and a sometimes ingenious choice of R1 and zener specs that will do the job. The worst-case extremes are high line voltage at minimum load (maximum zener current) and low line voltage at maximum load (zener tends to its knee region). The design problem is to choose operating conditions that will work. In on-off con-

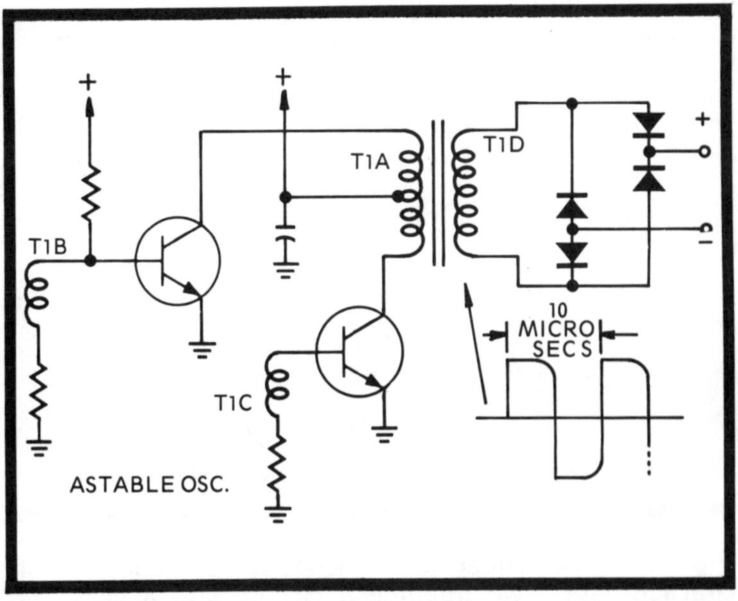

Fig. 4-15. A basic DC power converter with a typical sharp-edged signal appearing at the transformer secondary.

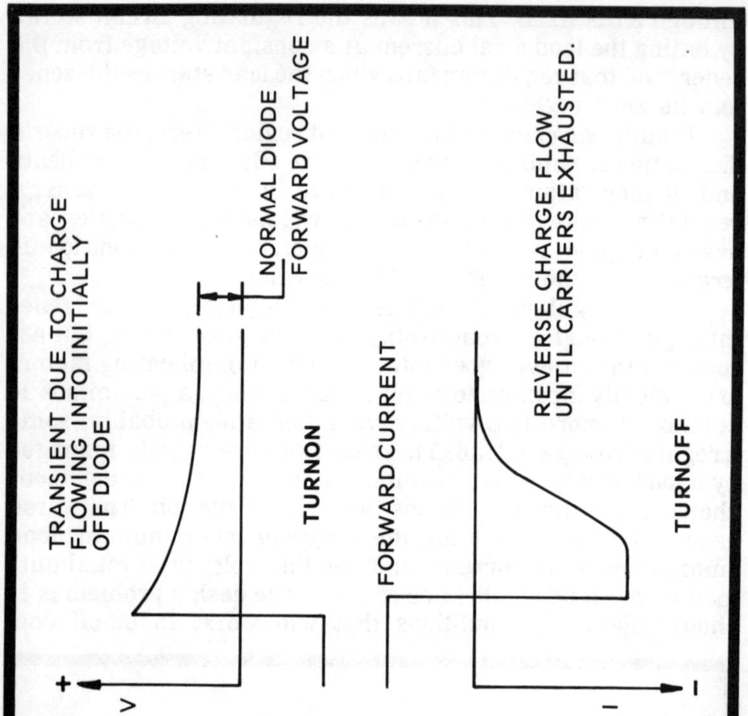

Fig. 4-16. To understand diode turnon and turnoff characteristics, we imagine a circuit with a very fast switch. Actual switching diode test units, like the Tektronix S unit oscilloscope adapter, work this way and produce curves like those shown here.

struction it helps to hand-pick a zener with a very low knee current, giving a wider permissible current variation.

DIODES AS FAST SWITCHES

The most important point about diodes as fast switches is that not all diodes are fast switches. Some of them are very slow switches, as such things go. Reviewing rectifier theory, we think of a diode as a switch closed if forward biased and open if reverse biased. What happens if it's slow and remains closed until reverse biased a little while? The simple picture of PN junctions says the electric charges responsible for diode conduction move too fast for any such problem to come up. In real life a slow diode can be a major problem—especially if you aren't familiar with the following:

For instance, you could be working on a DC-to-DC power converter in which transistors switch power in alternate directions through a transformer primary. See Fig. 4-15. Now, the higher the operating frequency the less core a transformer needs (each cycle transfers a dollop of power from primary to secondary; the more cycles per second the less core needed because the dollops are smaller) and so we have a circuit here working at 100 kHz. A 100-kHz sine or square wave spends 5 microseconds at one polarity, and the next 5 at the opposite polarity. However, a conventional power diode for 60-Hz use can take more than 5 microseconds to turn on or off. Evidently, if we use such diodes here they cannot keep up with the job and the supply won't work. If you think a diode is a diode you can be a very puzzled seeker after facts following several hours on this problem.

Some of the reasons real diodes work this way are very abstract, but one is plain common sense. When a diode is reverse biased its carriers are backed away from the PN junction in two very separate clouds. When we forward bias the diode, it takes time for the two clouds of charge to meet and for conduction to actually commence. See Fig. 4-16. And when we turn off the diode, at the moment of turnoff it is full of carriers (electrons and holes) which are quite indifferent as to which way they move; the diode conducts equally well either way for a brief instant. Computer switching diodes are engineered to minimize these effects; large power diodes for 60-Hz use show them very strongly. For a table of some suggested fast switching diodes see Fig. 4-17. In much amateur electronics work it's adequate to purchase a batch of "fast computer diodes" and, with a suitably suspicious state of mind, try them in their intended application. A practical example appears in Chapter 9.

TYPE	REVERSE VOLTAGE-MAX.	FORWARD CURRENT—MIN.	TURNOFF TIME
1N251	30V	75 MA	150 NS
1N456	25V	90 MA	
1N914	75V	75 MA	4 NS
1N4152	40V	20 MA	2 NS

Fig. 4-17. Some suggested switching diodes. Surplus diodes priced up to 25 per dollar are sometimes very good but should be used suspiciously.

DIODES AS GATES

Industrial and computer systems owe their tremendous speeds and utility to several factors, one of which is that we don't have to continually reach into their workings to flip switches. All such systems have in them, instead of switches, gate circuits. A "gate" suggests a gate-keeper, of course. Who tends the gates in computer and control systems? Circuits and signals numbering from one to a committee consensus! In this section we build on the basics of electric logic discussed in Chapter 3 to develop some concepts of gate circuits.

We begin with a brief discussion of circuit logic. Logic circuits are classed as series or parallel, positive or negative logic. The difference between series and parallel logic appears in Fig. 4-18. A parallel logic circuit (or system) has several outputs, with a single element of the overall output appearing at each. Every output may be associated with several inputs, and each box represents anything from one circuit to a system of them. But whatever is in the boxes, the output appears at a few terminals with a definite meaning assigned to each. For instance, if the parallel logic terminals here represent a 1, a 2 and a 4, the output shown reads: no 1, a 2, and a 4. This is a binary representation convertible to decimal 6.

The series logic circuit has just one output terminal. A string of pulses appearing on this terminal is the circuit's output. Upon seeing this string of pulses, we naturally wonder where the first pulse is; we discover this by looking at the gating pulse provided on another, nearby terminal. Any signal out of coincidence with the gating pulse is to be ignored; we observe that some of the output is timed inside the gating

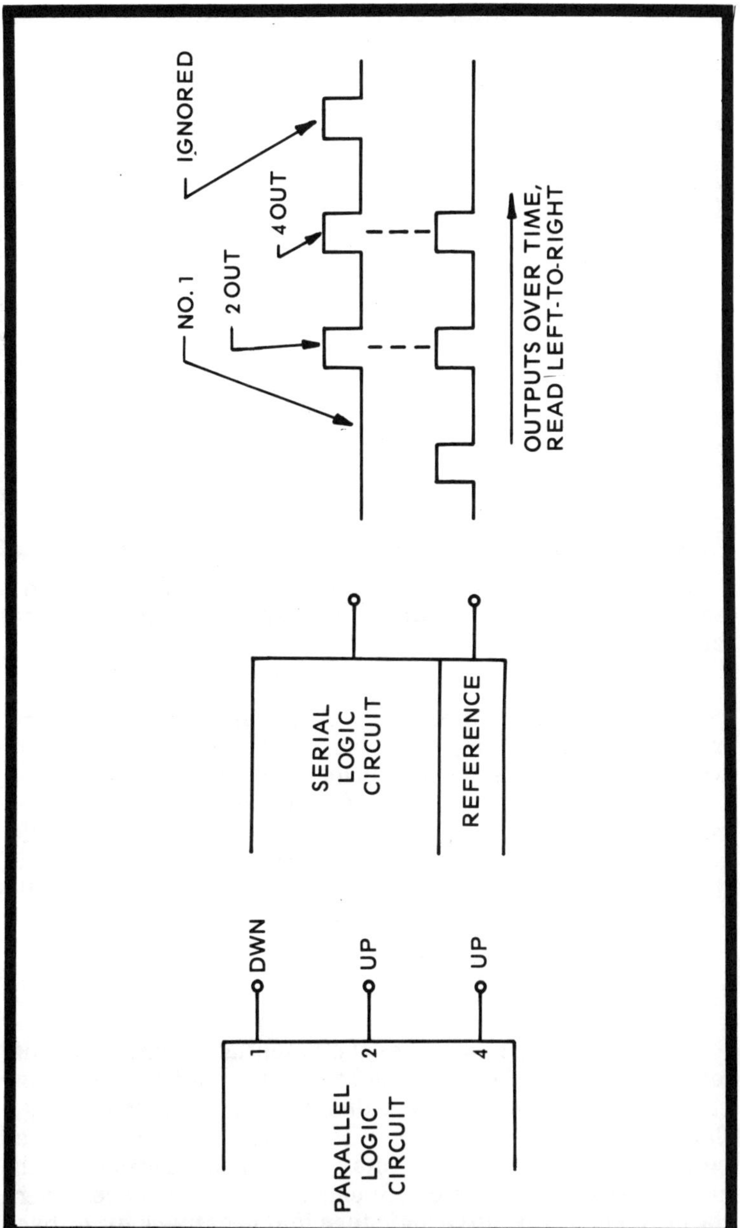

Fig. 4-18. Data carried by pulse and logic circuits is distributed over several terminals (parallel) or in time (serial).

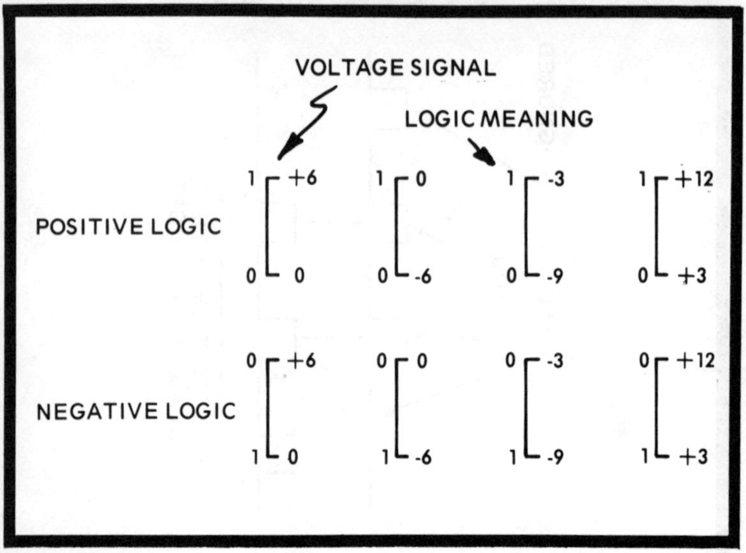

Fig. 4-19. Logic flags can be placed at critical points in a schematic or logic diagram as reminders of the local pulse coding.

pulse. Another circuit provides a clock pulse, and we interpret our output by observing no 1 pulse, but a 2 and a 4 pulse in step with the clock pulses. These add up to a decimal 6 again.

Neither series or parallel logic is inherently superior. Real control and computer circuits may work in either style; parallel is faster but series is cheaper (needs fewer parts). The same machine may use both, changing from one to the other by means of a "shift register."

Logic Flags

Only their utility matches the potential of gating circuits for generating confusion in the schematic reader's mind. One remedy is lots of practice. A valuable graphic aid is the "logic flag," a simple reminder of what voltage represents which binary value. One of these may be placed wherever in the circuit confusion is possible. Such key points are places where the designer lets positive logic turn into negative logic or back again, and places where logic voltages are different from elsewhere. In Fig. 4-19 we see some positive and negative logic flags. They are chosen to demonstrate that the circuit logic is the designer's choice, and it isn't immutable throughout the

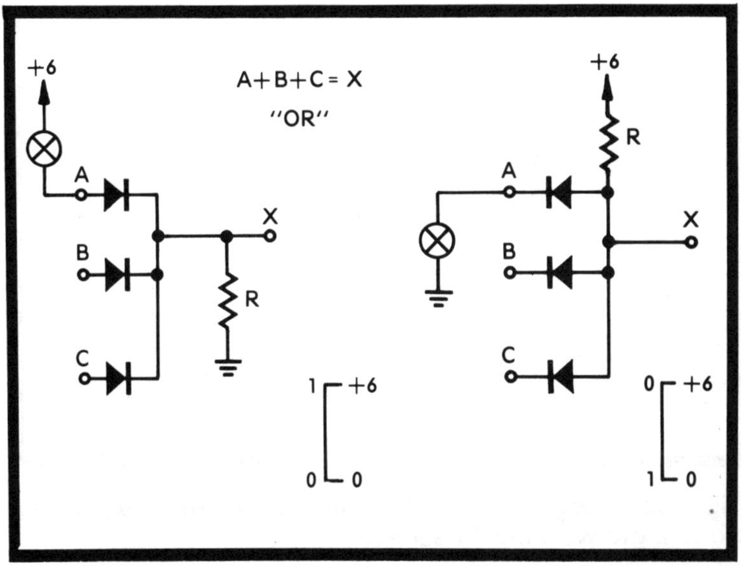

Fig. 4-20. Two simple OR gates, with Logic flags.

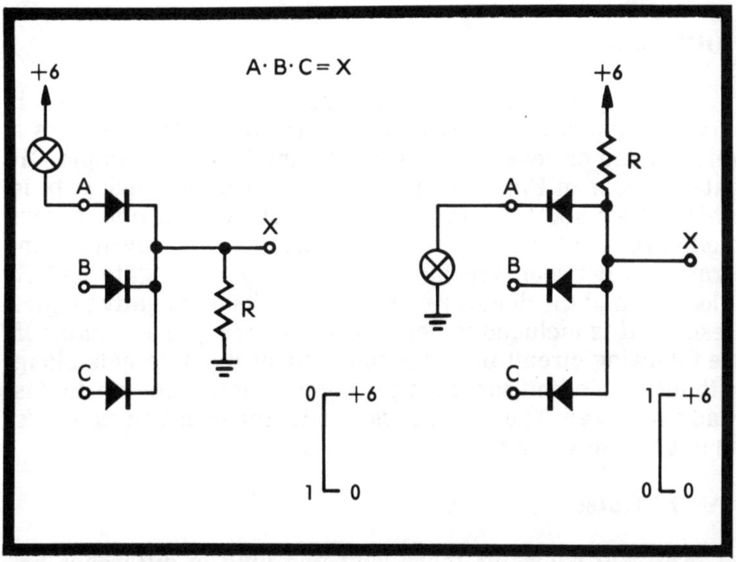

Fig. 4-21. Two simple AND gates, with logic flags. The circuits are the same as in the preceding diagram, only the flags are changed.

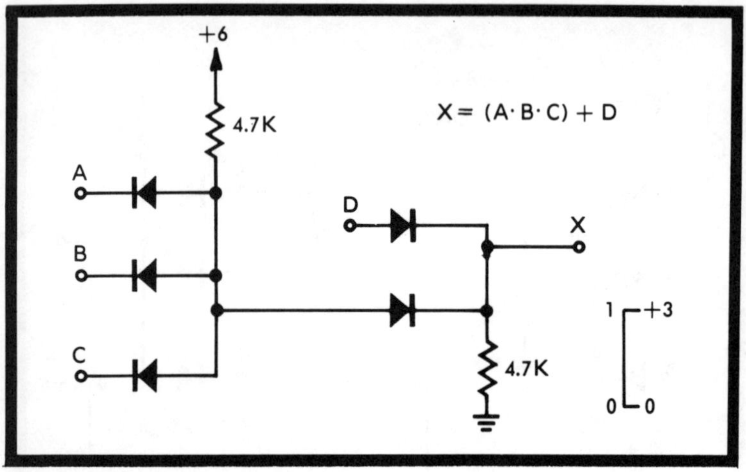

Fig. 4-22. "Two-level logic" circuits are simply two sets of diode logic working in series.

circuit. For instance, some logic circuits invert logic values and some do not, and actual logic levels may change considerably in passing through a series of circuits.

"OR" Gates

An OR gate is a basic diode circuit whose output is a 1 if any of its inputs are 1s. (An "exclusive-OR" gate generates a zero if none or several of its inputs are 1s.) Two simple OR gates appear in Fig. 4-20. The first is a positive logic gate in which, if we apply +6 volts at any input, we will find it (less diode drops) at the output. The diodes prevent unwanted interactions between the inputs. The second OR gate has R relocated and the diodes reversed; it works on negative logic. Resistor R is included to define current strength and path; if the following circuit provides this control the gate gets along without R. We find some simple algebra here, too, but this is read a new way. The "+" reads as OR; the equation says X is a 1 if A or B or C is a 1.

"AND" Gates

Fig. 4-21 shows us there isn't any built-in difference between AND gates and OR gates, since the circuits here are exactly the same as those of Fig. 4-20. Only the logic flags are changed, and that changes the circuits mathematical significance completely. Our positive logic OR gate is now a negative logic AND gate: its output is a 1 only if **all** its inputs are 1s, but the 1 is the same as a zero above. This is why we use logic flags. And the formerly negative logic OR gate is now a positive logic AND gate, again simply by reversing the logic polarity. The algebra for this circuit reads a "." as AND. It says we get out a logical 1 only when A and B and C are all 1s.

Two-Level Logic

To make up a more complex (and versatile) logical equation, we combine individual logic gates. See Fig. 4-22. We cannot go very far in this as long as we work with diodes, since a diode gate circuit output is far less powerful than any of the circuits feeding it. Usually, we can get in about two sets of gates: a gate feeding, say, one output terminal represents one level; several others feeding this one are the preceding level.

Kirchoff's Laws

The limitations of 2-level logic suggest this point, and it's important enough to stand a reminder. The performance and limits of diode gate circuits are keyed exactly to Kirchoff's laws. Several 4.7K resistors appear in the preceding illustrations; if they were omitted from working circuits some

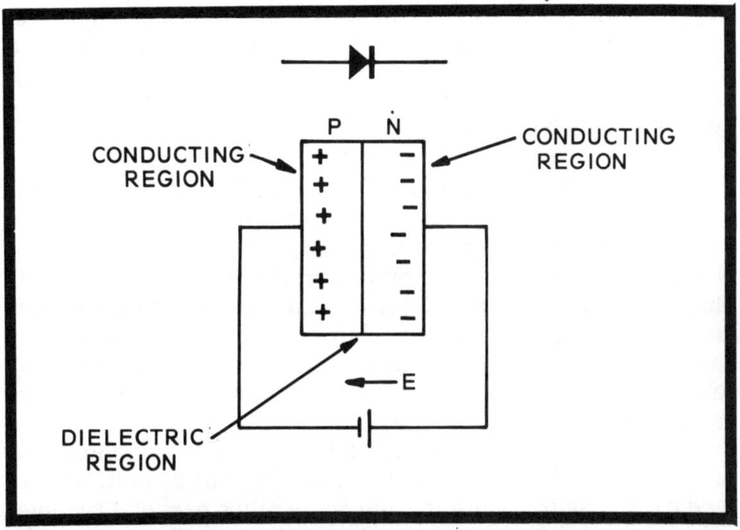

Fig. 4-23. A reverse-biased PN junction acts like a pair of capacitor plates with dielectric insulation between them.

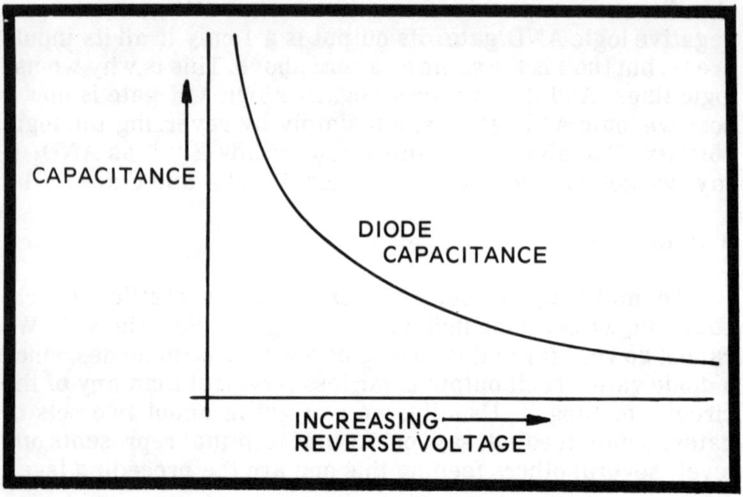

Fig. 4-24. Diode capacitance varies nonlinearly with reverse voltage. See manufacturer's specifications to learn the detailed characteristics of a diode capacitor.

current paths would be incomplete, and others would contain nothing to adjust current to a reasonable value. One wouldn't work positively (depending upon load) and the other would destroy diodes instantly. Diode gate circuits are a simple, powerful way to set up circuit logic; they are likely to be of central importance in the overall system. Don't let them be a central weak point.

DIODES AS CAPACITORS

Charge carriers in a reverse-biased PN junction retreat toward the terminals, leaving us with the situation shown in Fig. 4-23. Each charge carrier cloud resembles a capacitor plate, and the semiconductor material between the clouds is almost perfectly clear of electrons and holes. It becomes a fair dielectric material, and we wind up with a very workable capacitor. Its electrical value is varied by changing the reverse bias: the greater the bias the smaller the effective capacitance.

This opens the way to a host of practical applications. Since an electrical terminal tunes a resonant circuit, we can feed back a control voltage from some other part of a receiver system, making the receiver self-tuning. The automatic-frequency-control circuit of an FM receiver usually works this way, for instance. And the voltage-variable capacitance effect can be used as a very efficient frequency multipler, putting out several watts of radio-frequency power at two or three times the input frequency. This is an important technique at the hundred MHz and higher frequencies. Good diode capacitors are now very popular in consumer electronics equipment and have become available at low prices. For small-signal use they are superior replacements for the clumsier, electrically less efficient mechanical variable capacitors.

The curve of Fig. 4-24 shows how capacitance depends on reverse voltage for a typical diode capacitor. The relatively high voltages needed for maximum tuning range are little disadvantage for many applications, which may require

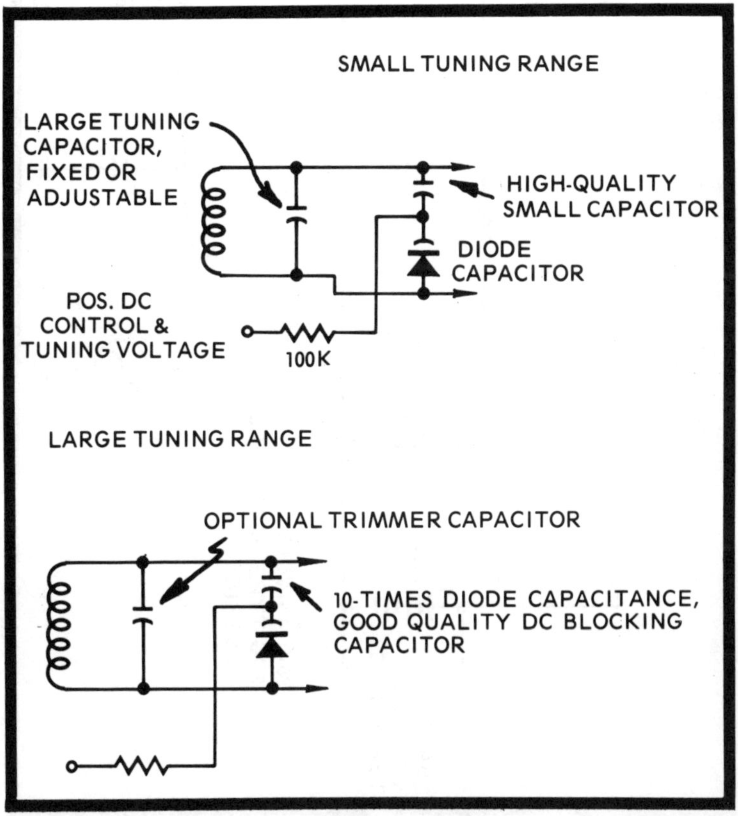

Fig. 4-25. Diode capacitors tune circuits like mechanical variable capacitors, but must often be installed with a blocking capacitor. Voltage swing in the tuned circuit should be moderate compared to the diode bias voltage.

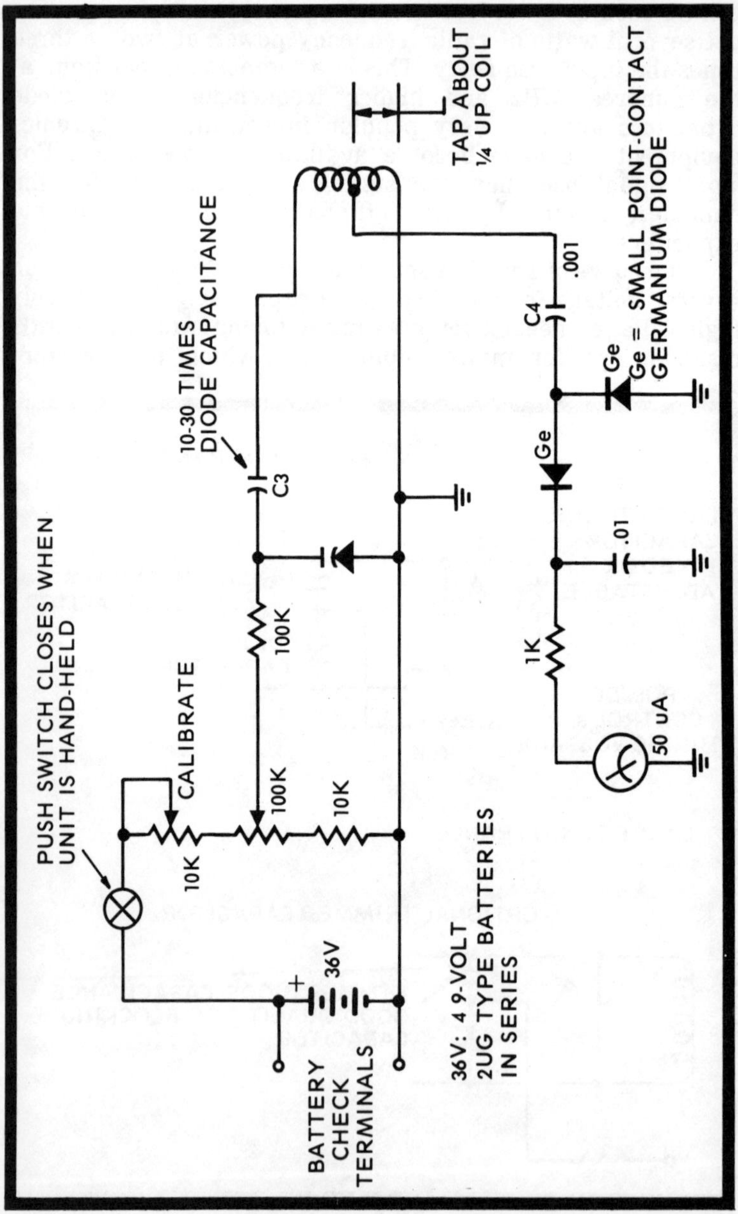

Fig. 4-26. An elementary but very practical diode capacitor application is found in this absorption wavemeter. Diode capacitors are good in high-frequency circuits because they are physically small and solid.

tuning over a percent or less of the operating frequency. And since the diode capacitor draws practically zero current, inexpensive batteries will show good stability. For instance, a set of 2U6 type 9-volt batteries would work well, delivering a tenth to one milliampere into a potentiometer tuning circuit.

Diode capacitors don't have to be specially made units. The base-collector diodes of high-frequency transistors are good prospects for small diode capacitors. The high-quality low-priced epoxy transistors manufactured by General Electric look like excellent prospects for diode capacitor application and should offer up to 20 or 30 pf at low reverse voltage. Some modern rectifier diodes could turn out to be good capacitors, but older types are markedly unstable.

A diode capacitor's control voltage must not be lost in the tuning inductance. A blocking capacitor is needed for most circuits, but may be avoided for some high-frequency cavity arrangements. See Fig. 4-25. The capacitor need not be very high quality if its reactance is small compared to that of the tuning capacitor. A practical minimum is ten times tuning capacitance; if the diode capacitor is 25 pf maximum, then 250 to 1000 pf will make a good blocking capacitor choice.

A very small but high quality blocking capacitor can limit the tuning range where a large range isn't needed. A parallel capacitor makes up the rest of the tuning capacitance. This also minimizes the unfortunate drift effects found in some diodes not meant for tuning capacitor work. For instance, a silicon power diode of very poor stability, wired in series with about a 3 pf good capacitor, worked very well sweeping a 25-MHz oscillator over about 5 kHz.

Fig. 4-26 depicts a very simple diode capacitor application. This is an absorption wavemeter, with the diode capacitor serving as the entire tuning capacitance. Parts values aren't critical, and the finished circuit is calibrated by comparison with known signal sources. Notice C3 and C4 are chosen for the desired frequency range as having a reactance of roughly 100 ohms at the operating frequency. A nice way to construct this wavemeter is to use one tuning head for all frequencies below 75-100 MHz, and to build separate tuning heads for higher frequencies. The tuning dial is calibrated 0-100 and simple charts or lists relate the dial setting to frequency.

DIODES AS PROTECTIVE DEVICES

A few well-placed diodes apparently not doing anything in an electronic circuit is usually an indication that its designer was a careful and alert engineer. He puts the extra diodes

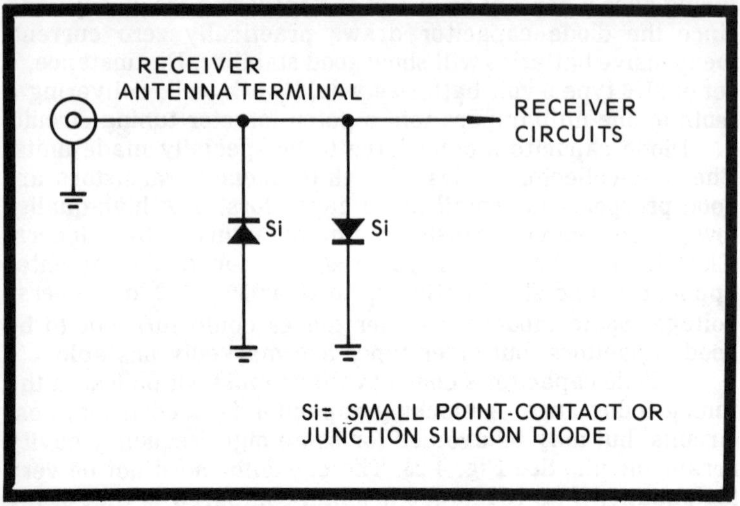

Fig. 4-27. Diodes are easily installed to prevent receiver circuit damage by too strong radio signals.

where they don't affect circuit operation—until something goes wrong. Then the diodes correct a bad situation by shunting, absorbing, or regulating a harmful voltage or signal.

For instance, a typical protective diode application seen in radio receivers appears in Fig. 4-27. Impedance levels in this circuit tend around 50 to 75 ohms, so the tiny capacitance of a pair of point-contact diodes does not affect circuit operation. But if signal levels exceed roughly 0.2 volts for germanium diodes, the diodes go into conduction. A bad overload may spoil the diodes, but these are easily replaced without making any circuit adjustments. Typical signal levels in receiver front-end circuitry are in the 100 microvolts and down region; if the receiver is very near a transmitter with signals tending into the 1000 microvolts and up region this scheme cannot be used. The diodes now act as signal mixers, introducing very bad intermodulation interference.

A discharging capacitor may deliver a considerable charge into a transistor's base-emitter diode, warming it considerably. Some transistor structures cannot endure this treatment. They go into zener breakdown at under 5 volts and fail quickly from overheating. Adding a protective diode stops this perfectly. See Fig. 4-28. As soon as the base-emitter diode sees above 0.7 volts reverse bias (evidently a silicon diode here) the protective diode goes into conduction and prevents a higher voltage from appearing across the diode. This

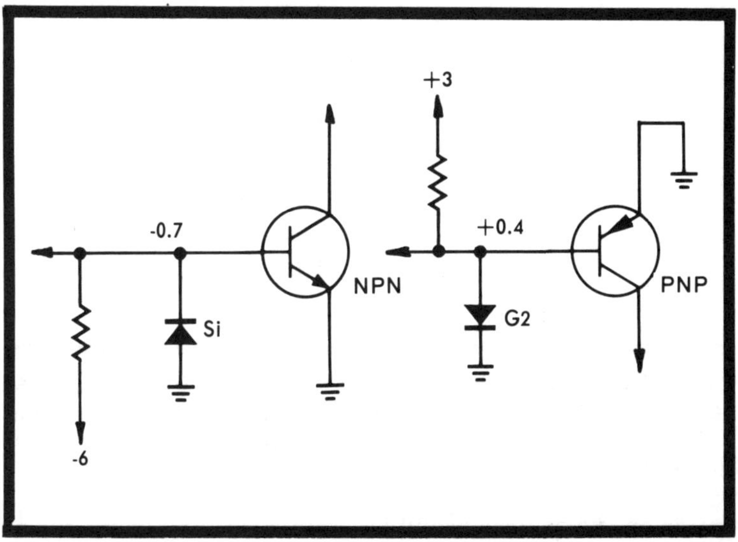

Fig. 4-28. This simple provision bypasses excessive reverse bias or signals to the transistor base-emitter diodes.

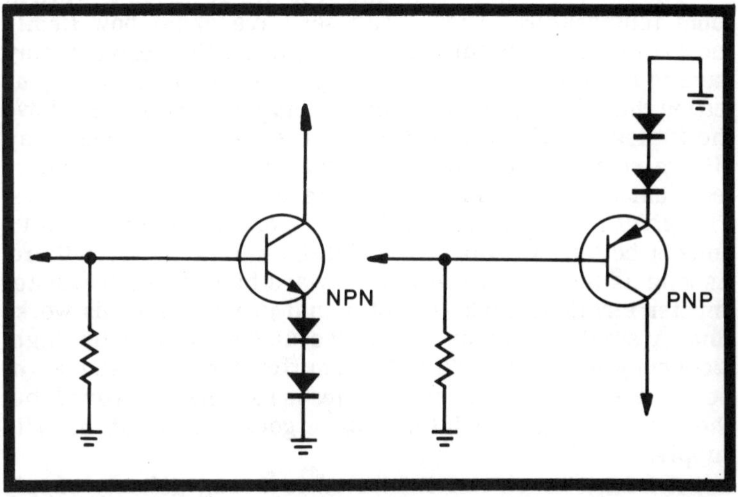

Fig. 4-29. If a transistor emitter terminal is held a few hundred millivolts off ground, this voltage can be referred into its base circuit to turn the transistor very firmly off.

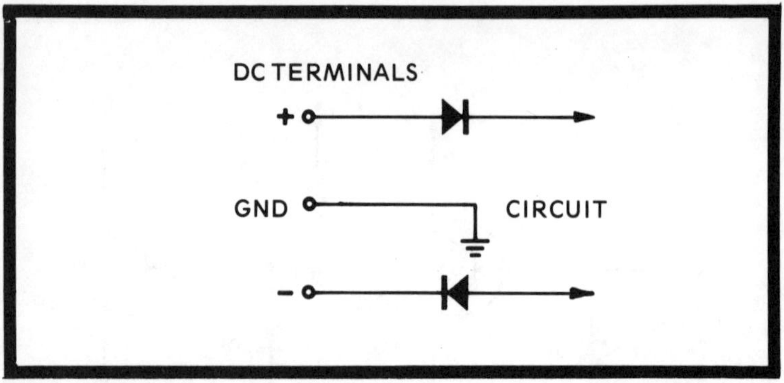

Fig. 4-30. One diode in a supply lead protects against reverse-polarity connections. This is better than a fuse which may blow too slowly to protect fragile semiconductors.

provision is also very useful in circuits working from bipolar supplies, where one supply may come on well before the other. Wrong voltages and currents continue until the other supply comes on, and well-placed diodes can absorb much of the resulting abuse. (Notice to designers: design any circuit for bipolar voltage operation to survive indefinitely with either power supply off.)

The switching circuits in this book are designed to work at room temperature. They aren't sensitive about how firmly their transistors are turned off. In real life the designer may have to compromise between using a second power supply or letting the off state get a bit soft as temperature passes above the 30 degrees C level. A couple of well-placed diodes may offer a remedy. See Fig. 4-29. Here the transistor's emitter terminal is off ground by the voltage across one or two diodes. This gives the designer 0.2 volts (one germanium diode) to 1.4 volts or better (two silicons) and it doesn't have to stop there. As long as the diodes are well forward biased, the transistor emitter terminal thinks it's at signal ground and this works fine. A small capacitor across the diodes is good for high-frequency operation. With the emitter terminal above (or below) ground the base bias resistor is returned to ground, but the circuit now acts as if the resistor goes to a turnoff polarity supply.

A transistor circuit connected to wrong polarity supply voltages is in trouble immediately, if not instantly. Diodes in the supply lines can stop wrong-way charge flow. See Fig. 4-30. If zener regulators are used, this circuit can be improved further since the zener acts like an ordinary diode when forward biased. (It normally works in reverse breakdown.) See Fig. 4-31.

Here is an example drawn from real, live, hard experience. The box labelled "electronic chassis" in Fig. 4-32 contained over 50 transistors and associated components, all protected as described above. It was destroyed by a user who knew nothing of circuits. How did he do it? He hooked it up—without a ground return line. Can you redesign the supply lines circuitry so that the system cannot be destroyed again?

DIODES AS DETECTORS

"Detector," as the term appears here, is an old-fashioned word for something that converts a radio-frequency signal into an audio-frequency signal. Our less knowledgeable predecessors applied it to a variety of distantly related processes. These fall under the general headings of large-signal detectors, small-signal detectors, and what we now commonly call "mixing."

Large-Signal Detectors

The data carried by an amplitude-modulated radio signal is coded into it by fast variations in its amplitude. See Fig. 4-33. One way to get the data out again is simply to amplify the signal to a high level, compared to typical diode working voltages, and recitfy it. For this approach the signal is built up

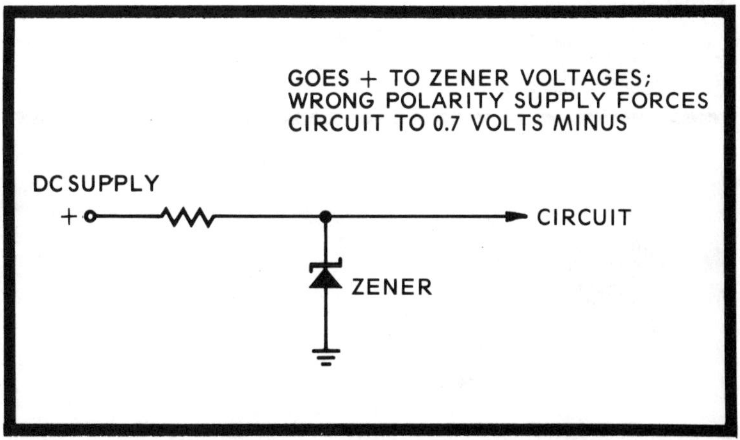

Fig. 4-31. A zener diode also protects against supply connection polarity reversal, since a reverse-biased zener turns out to be a forward-biased silicon diode.

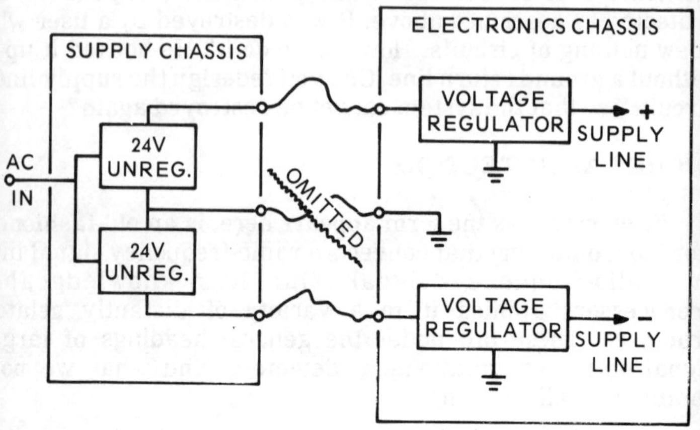

Fig. 4-32. Here is a sure way to burn up valuable electronic equipment. What provision prevents this error, if repeated, from destroying semiconductors? (Anything goes, if it works.)

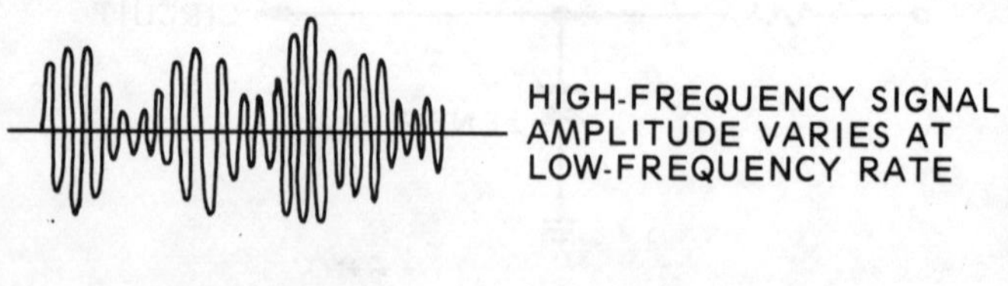

Fig. 4-33. This waveform shows how information is coded into amplitude-modulated radio signals.

from its incoming value of a few microvolts to millivolts. Usually the receiver's IF (intermediate frequency amplifier) does this.

Then the signal is rectified into a small capacitor circuit that is filtered and loaded enough for the voltage across the capacitor to follow fast variations in signal strength. See Fig. 4-34. There is lots of signal power, and some typically goes to an additional filter that smooths out the audio variations, leaving an average voltage to feed back to preceding stages. This DC-like return signal is called the AGC voltage (automatic gain control) and continually adjusts the receiver's sensitivity to compensate for changes in received signal strength. No large-signal detector application circuits appear in this book.

Small-Signal Detectors

When signal strength is comparable to or smaller than the typical diode operating voltage, we pick a germanium diode as detector and use its square-law transfer characteristic. This kind of detection is well described as "very sloppy rectification." See Fig. 4-35. Since the germanium diode thinks in hundred-millivolt terms and the incoming RF signal is maybe

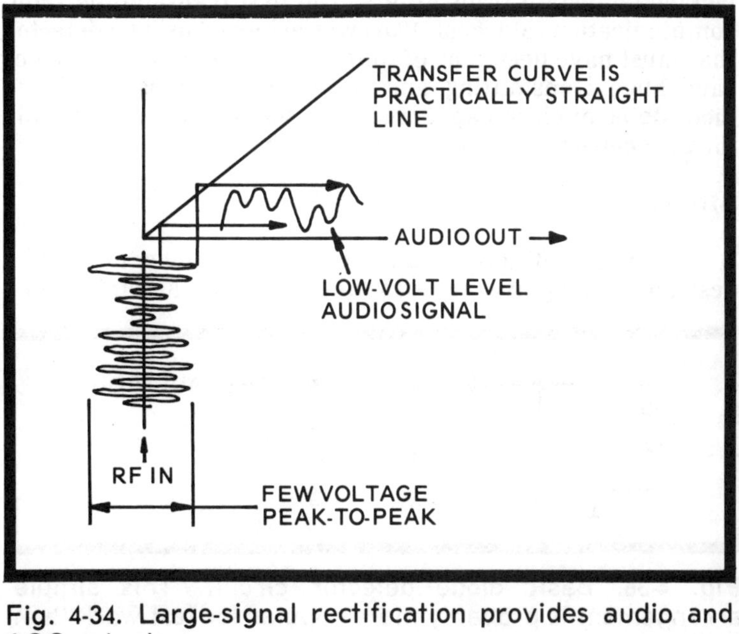

Fig. 4-34. Large-signal rectification provides audio and AGC outputs.

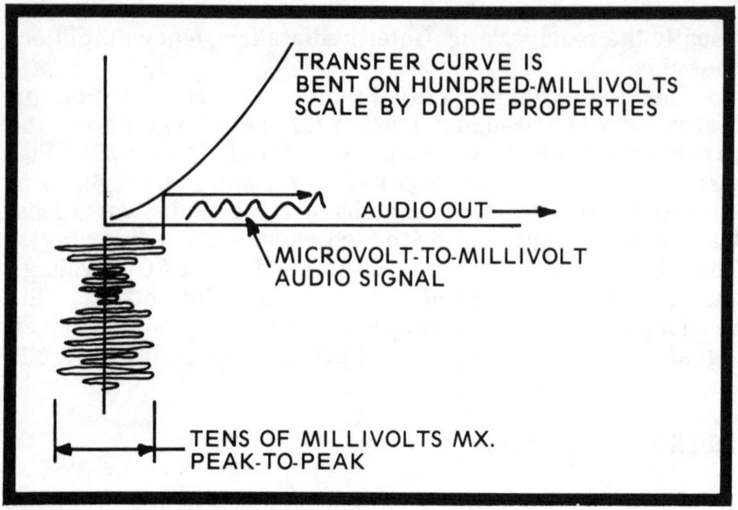

Fig. 4-35. Small-signal rectification is far less effective since a smaller part of the diode curve is used.

1 millivolt, small-signal detection is very inefficient. Yet it works well enough, and many records have been achieved in picking up distant transmitters. The first transatlantic radio communication (Marconi, 1901) was achieved using a detector that must have been tens of times poorer than today's diodes (and huge transmitting and receiving antennas). A basic diode detector circuit appears in Fig. 4-36 and a very workable diode receiver is described in Chapter 6.

Mixers

When two different radio-frequency signals meet in a resistor, nothing new appears. But if they are assembled in a

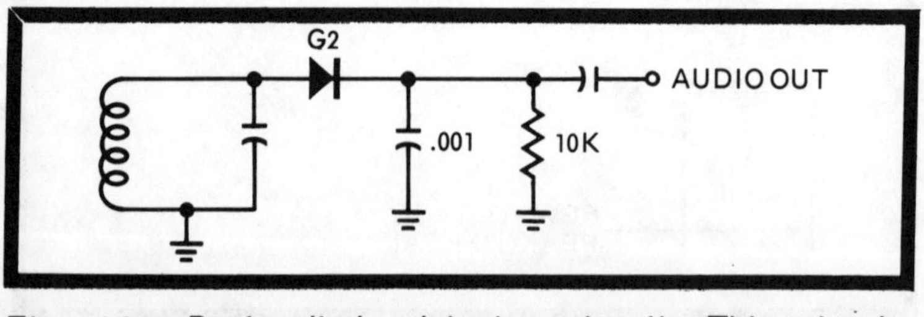

Fig. 4-36. Basic diode detector circuit. This simple arrangement is probably more popular for test work than radio receivers.

diode circuit they are "mixed" and new frequencies appear. One new frequency is the sum of the incoming frequencies; another is their difference. Additional frequencies appear, but they are of no concern here. There is no limit on the frequencies generated, and frequency mixing is commonly used to generate an audio frequency output from a radio frequency signal.

An elementary, convenient, and highly useful diode mixing circuit appears in Fig. 4-37. The RF signal sources are termed "local oscillator" and "RF input," since this is what they are most likely to be. Two pickup links wired in series present the voltage sum of these signals to the diode, which generates the new frequency sum and difference signals. The entire mess is then presented to an output tuned circuit, if it is radio frequency, or to an audio pickoff circuit, if audio. Capacitor C is chosen for low impedance (say, 10 ohms) at the lowest frequency in the circuit and resistor R simply indicates good oscillator coupling by converting the diode current into a voltage reading. About 1 volt across R is a good trial value, and the oscillator and RF input coupling are later adjusted for best receiver performance.

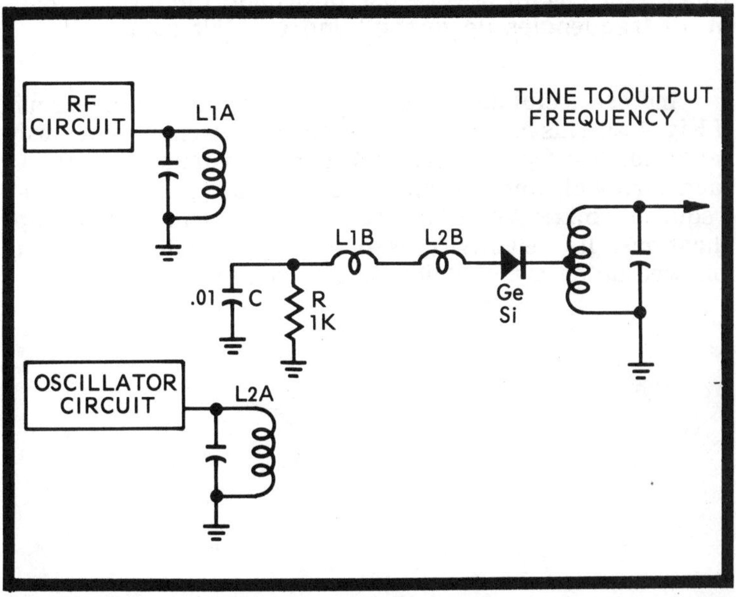

Fig. 4-37. Signals can be mixed in diode circuits by series addition from low-impedance loops. A tuned circuit must be used in the output to de-emphasize the unwanted input frequencies.

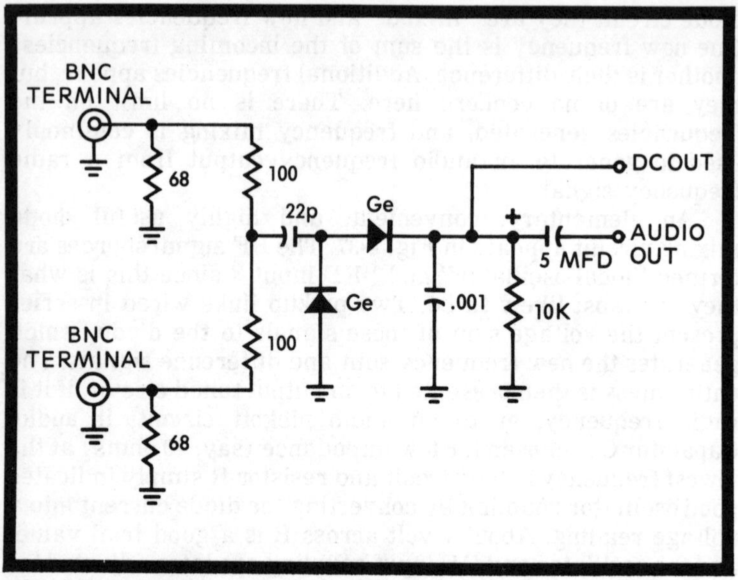

Fig. 4-38. Signals can be added by parallel circuits too. This mixer circuit is very useful around the lab for tests on all frequencies up to the hundred-MHz range.

Another useful diode mixer is the resistive adding circuit of Fig. 4-38. This is inefficient for receiver use, but fabulously handy for test work. About a 50-ohm resistance appears at each terminal, and the circuit can be used as a detector, frequency mixer, dummy load, and for many other applications. Its sensitivity is quite remarkable when it is followed up by a high-gain audio amplifier.

Chapter 5

Switching Transistors

A diode is basically a switch working according to the rule: it is closed if forward biased and open if reverse biased. On or off, in other words. If we view transistors in the same general way we find their characteristics are relatively simple. And we keep our thinking right down to brass tacks since transistors are often used as switches. Computers, counters, and control circuits, all use large numbers of transistors as switches (maybe hundreds of thousands in a large machine) and the extremely high reliability of these machines is in large part because the transistors are working as switches.

Chapter 4's diode discussion shows the utility of these simple devices, and it shows something of their surprisingly complicated behavior. A transistor, it turns out, is basically a pair of diodes built very close together, interacting within one tiny piece of semiconductor crystal. This interaction and its consequences mask diode behavior, giving an appearance of a new and complicated electronic device. A very nice way to start understanding how transistors work is to begin with their simplest way of working; that is, transistors as switches.

It turns out that you need very little special knowledge and experience to begin designing transistor switching circuits and systems. In fact, maybe you will find switching transistors and their circuits so challenging and interesting, like a fascinating game, that you'll never try linear amplifier and radio circuits. If your interests turn that way, your next step, after Chapters 5 and 9 in this book, would be learning more about digital IC technology.

SATURATION AND HARD-OFF

Transistors can switch considerable amounts of power, yet they are fragile. Cut or break open a "transistor" and look inside for its real electronic structure. This turns out to be nearly or actually microscopically small. The real transistor cannot endure temperatures far above the boiling point of water, and because of its small size, very little energy is

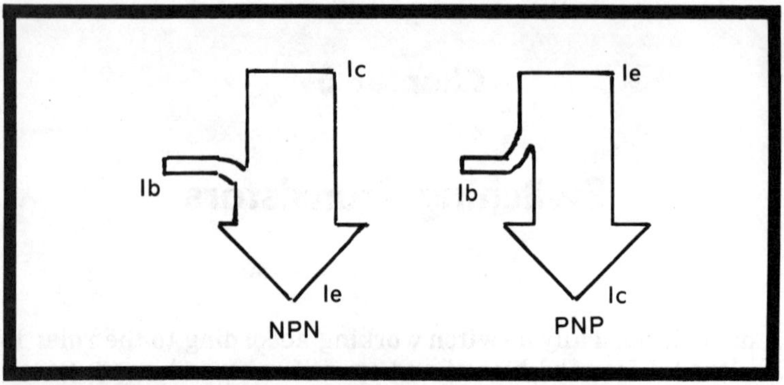

Fig. 5-1. Arrow widths are proportional to the charge flow through each transistor terminal. This arrow image is a nice way to keep clear the difference between PNP and NPN transistors.

needed to overheat it. Its designer expended much ingenuity combining good electrical properties with a really effective thermal connection to the large outside case. Yet, however good a job he did, your circuit design must never ignore the real transistor's delicate constitution.

There are two operating regions where a transistor's delicacy is relatively unimportant. These are "saturation"—high current and low voltage; and "hard-off"—high voltage but practically zero current. A transistor's power dissipation, like a resistor's, depends upon voltage across times current through its structure. In saturation or hard-off, its power dissipation is near zero since the V x I product is small. A one-tenth watt transistor might switch ampere current pulses at 12 volts or more. This service looks right to cook it instantly, and yet it survives because it never overheats. Exactly how the circuit designer manages this is interesting and highly useful information.

Transistor Beta

In Chapter 3 we covered the basic ideas of transistor operation. Charge flows in the base-emitter diode and most of it is stolen by the reverse-biased base-collector diode. This basic concept is very clear for NPN and PNP types if we illustrate it by two arrows as in Fig. 5-1. The arrow widths represent the amount of charge flowing. The narrow entering or exiting arrows at the left stand for base current. These drawings show that emitter current equals the sum of base and collector currents. If we decrease base current to zero, collector current also goes to zero. And whatever the base current, collector current remains larger by about the same factor. This factor is called transistor "beta."

A transistor's beta is simply its natural ratio of collector-to-base current. If 1 milliampere of base current invokes 50 of collector current, the transistor's beta is 50. If 20 microamps of base current invokes 20 milliamps of collector current, its beta is 1,000 and there are units with this very high current gain. General Electric's 2N5305 Darlington transistor has a nominal beta range of 2,000 to 20,000. Since transistor beta is simply collector current divided by base current, we can measure it easily. See Fig. 5-2.

This circuit can be assembled on a piece of perfboard or it might even be manufactured "frozen style" as a test instrument. Since the base-emitter diode is forward-biased its voltage varies, depending upon transistor type, from perhaps 0.2 to 0.8 volts. This is less than 10 percent variation, compared to the supply voltage. It's nice to test either germanium or silicon transistors here, so we suppose this voltage is 0.5 volts. That isn't far off in either particular case. We assume 11.5 volts from the battery (allowing for a bit of battery aging) and come up with the bias resistors shown. A multimeter plugged into the collector circuit reads the resulting collector current. Since base currents are chosen as ten multiples, we can work out transistor beta without pencil and paper. The series RC connections from collector to base and base to emitter kill the transistor's high-frequency response, preventing oscillation of

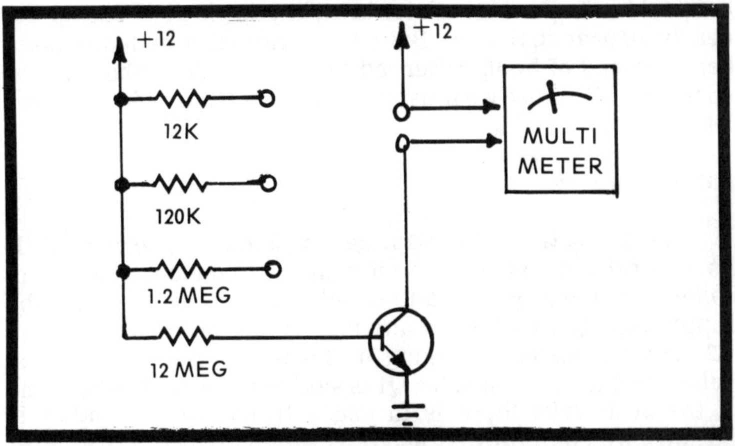

Fig. 5-2. This circuit measures a transistor's beta by providing a small base current and observing the resulting collector current.

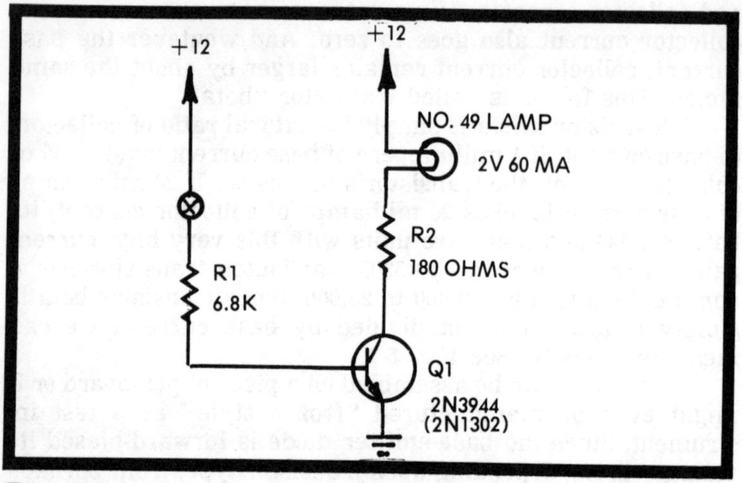

Fig. 5-3. In this circuit a 2N3944 transistor (it could be a 2N1302) is acting as a switch. The design ignores collector-to-base leakage, which is no problem at room temperature under these easy operating loads.

some highly active modern types. They have no effect upon DC beta tests.

This circuit could be modified by adding reversing switches so it works for both NPN (shown version) or PNP transistors. It ignores collector-base leakage, which is discussed later. (Can you do something about that?) It yields only ballpart readings of transistor beta, but this is good enough for nearly all practical work. But why is transistor beta discussed here, instead of being reserved for the chapters about linear operation? Because we must know what it is, in order to never let the circuit use it!

Saturation

In Fig. 5-3, transistor Q1 is getting 5 ma of base current. If its natural beta is 50, it would like to conduct 250 ma of collector current. But it can get only 60 ma or so, because the supply voltage won't move any more than that through L and R2. The transistor's working or **forced** beta here is 12, far below its natural beta of 50. Q1 is said to be in saturation—one of the best ways there is to use a transistor (the other is turnoff; we come to that soon).

When a transistor is in saturation, its collector current is limited by outside components, not by the transistor. Usually, the outside components are more stable and reliable than the transistor. In saturation the transistor's collector voltage is roughly one-half volt. Other transistors plugged in its place cannot depart far from this (if the circuit biases them into saturation) even if they are silicon types in a germanium circuit. And at this low voltage, a transistor carries a heavy current before overheating becomes important. If the transistor is working at a forced beta of 12, then it makes no practical difference if its natural beta is 20 or 200, since this is no factor in circuit operation. All this means that saturating transistor switching circuits are very nice to work with.

Two key design points appear now. Fig. 5-3 is an indicator working at moderate current levels. Some saturated transistor circuits manage heavy surges; in these applications the manufacturer's opinions regarding maximum transistor current deserve attention. A saturated transistor can be perfectly happy at a current load that melts one of its inside wires (these are sometimes too small to see). And the other key point is that the transistor must be guaranteed to stay in saturation. For instance, a transistor carrying 500 ma pulses might read an average of 10 ma, but a base signal adequate for 10 ma saturation could be far too small for saturation during the pulses. If a saturated switching transistor perishes a few moments after its circuit is turned on, probably it isn't getting enough base current during part or all of the operating cycle. Switching transistor forced beta operation must never be allowed to approach natural beta.

Hard-Off

The extreme opposite of saturation is "hard-off," with the transistor now resembling an open switch. This may be hard to achieve for some germanium transistors or for silicon transistors working at warm temperatures. All transistors have a built-in leakage which tends to turn them on. We have ignored this until now, and it can even be handled lightly in some critical circuits. But if we want to turn our transistors really hard-off, we must give a thought to this leakage current.

In Fig. 5-4, we have an equivalent circuit whose inside transistor is assumed perfect. The outside ring represents the transistor as the designer and circuit see it. In between, there is a Norton generator (from collector to base). This represents the leakage source (Icbo current), which is simply the same diode reverse leakage mentioned in Chapter 4. Just like the rest of diode theory applied to transistors, this comes along too. It's an unwelcome guest.

The Norton generator leaks a few microamps at room temperature for germanium, typically a few nanoamps for

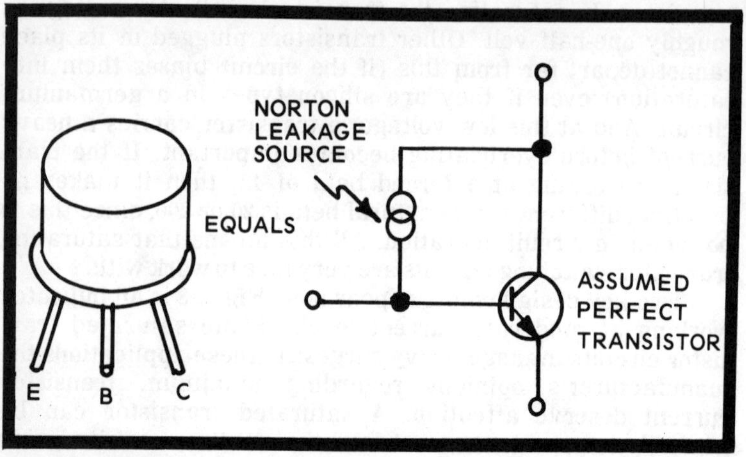

Fig. 5-4. Leakage cannot always be ignored. Here we imagine that a Norton generator connected from collector to base of an NPN transistor accounts for the leakage.

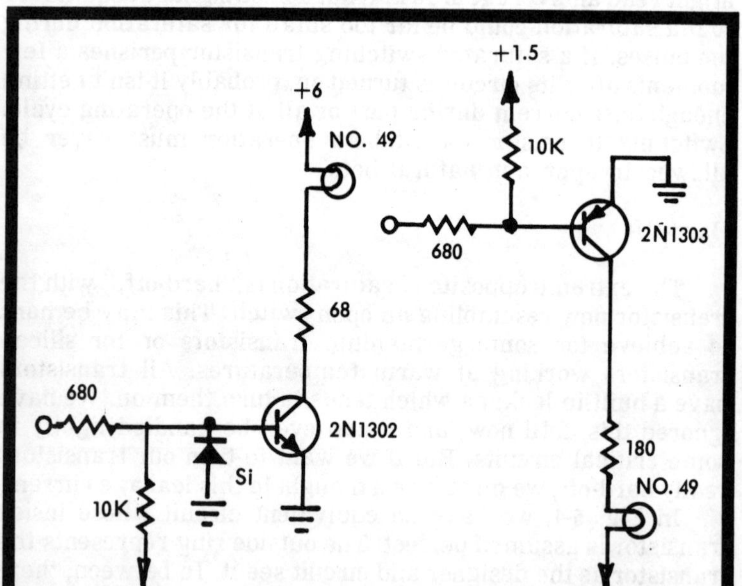

Fig. 5-5. A transistor is turned hard-off when its base-emitter diode is a few tenths volt reverse biased. The turnoff bias can be provided from a low-voltage source or a higher-voltage source with added diode protection.

silicon transistors. A large germanium power transistor may leak many milliamps. As the transistor is warmed, leakage rises, and it all goes to the transistor base terminal. At high temperatures it may be enough to urge the transistor near conduction or even hard-on, upsetting circuit operation and perhaps leading to destruction of the transistor. How can we defeat this leakage current? Simply by letting it run out again. If we have 10 microamps leakage into the base terminal, we can add a connection that leaks 10 microamps out. Things aren't really this simple. Leakage increases sharply with temperature. We have to make circuits that overdo the job at room temperature in order to work at high temperatures. See Fig. 5-5 in which we turn off our transistors from an extra power supply. We can choose a low-voltage bias supply and reverse bias the base terminal directly, or we can pick a higher voltage source. Then we must arrange not to reverse "overvolt" the transistor base-emitter diode. One or two protective diodes, in conduction when the transistor is off, do this very nicely.

Leakage protection weakens when leakage overpowers the turnoff bias circuit. A good rule of thumb that hard-off is softening is evident when the transistor base terminal comes to zero volts with respect to the emitter. In Fig. 5-5 this takes about 150 microamps in the first case and 1.2 milliamps in the second. Good design limits this turnoff bias to what the circuit really needs, since the turnon signal must defeat it in normal operation.

THE ASTABLE

What circuit switches a switching circuit? Usually the chain of control winds up in the outside world as some connection to a switch, a readout or readin device, lamp or relay—the list is almost indefinitely long. But some switching circuits haven't any outside control at all. The astable oscillator is one of these, where "astable" means "without resting place." The astable's purpose in life is to turn itself off—but it never finishes the job. Astable applications include generating gate, timing and square wave, and harmonic-rich signals, pulse generation, lamp flashing, etc. Basically, astable operation is simple, with the most observable aspect being an on state that seems to jump back and forth between two transistors.

As the engineer sees things, the on state jumps because it's pushed by an off state. In Fig. 5-6 we have a simple, tested astable circuit that flashes a pair of lamps slowly enough to observe its sequence of events. To understand this circuit we

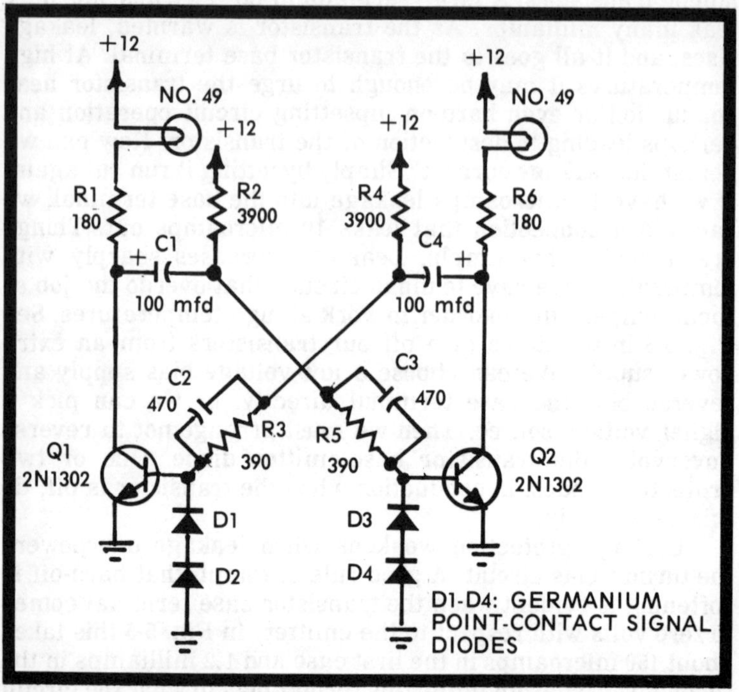

Fig. 5-6. This lamp-flashing astable could as well generate a timing signal for other electronic circuits. Large capacitors C1 and C4 control the flashing rate.

choose a particular point in its never-ending cycle of operation and proceed from there. We suppose the on state just jumped to Q1 and, as a reminder, we sketch three or four diagonal lines inside Q1 to help us remember it is in saturation, conducting heavily.

At this instant L1 is lighted and about 50 ma is flowing down through it and R1 into the Q1 collector. The Q1 collector voltage fell from 10 volts to 0.2 volts when Q1 turned on, and this negative voltage step appears through R5 and C3 at the Q2 base as a turnoff signal. Since Q2 has just turned off, its collector voltage is rising to feed a turnon signal back to the Q1 base, but since the transition is finished very little signal is feeding through C4 back to the Q1 base. Most of the Q1 base current is coming down through R4 and R3, which are chosen to deliver enough current to hold Q1 in its present saturated state. At this point it looks like Q1 could rest indefinitely in saturation since its base and collector terminals are connected through appropriate resistors to adequate voltage to maintain saturation indefinitely. But events elsewhere will upset this soon; the astable never rests.

When Q1 pulled the C1 terminal about 10 volts groundward, C1's other terminal followed and came momentarily to rest near -1.0 volt. It could have gone further, but protective diodes D3 and D4 went into conduction to limit the turnoff voltage at the Q2 base. Therefore, no charge is flowing into the Q2 base. The upper end of R2 is at 12 volts and its lower end near -1.0 volt, so charge is flowing down and into C1, allowing the Q2 base terminal to settle up toward and finally above ground level. About the time the Q2 base gets around +0.1 or +0.2 volt above ground, things begin to happen.

Q2 begins to conduct. This starts bringing down its collector voltage. C4 feeds this negative-going swing through R3 and C2 to the Q1 base, which sees this as a turnoff signal. As Q1 starts going off its collector voltage rises, and this rise goes through C1 and arrives at the Q2 base as a turnon signal. This feedback from Q2 to Q1 and return escalates rapidly and the on state seems to jump from Q1 to Q2. (For the quite slow 2N1304 transistors used in the original breadboard, transition took about 2 microseconds.) Now Q1 is hard-off and Q2 is in saturation, as indicated by a lighted L2. But now C4 is losing its stored charged as current leaks through R4 and we are right where we came in, except Q2 is on rather than Q1.

This can go on indefinitely, but we must examine details. First, the small capacitors, C2 and C3, simply feed control signals directly from one transistor's collector to the other's base terminal. They are called "speed-up capacitors," and often appear in switching circuits. This astable hardly needs them, but a scope check will show stronger switching if they are included. When each transistor goes on, the timing capacitor (C1 and C4) in its collector circuit can deliver a large current through the transistor collector, to ground, and back to the capacitor through the other transistor's protective diodes. This current could hold the on-going transistor out of saturation. Resistors R3 and R4 limit this current so the on-going transistor saturates quickly. We have already mentioned the protective diodes, which aren't always necessary but are included here for a better picture of a complete real circuit.

Much simpler astables appear in many magazine articles. Many work fine, but sometimes one of them (and maybe even this one) won't start. Why is this? As we look at the schematic, we see Q1 and Q2 are biased, on the average, hard-on. When power is applied, if both happen to lock on, then neither can turn off the other. To prevent any chance of this lockup we add

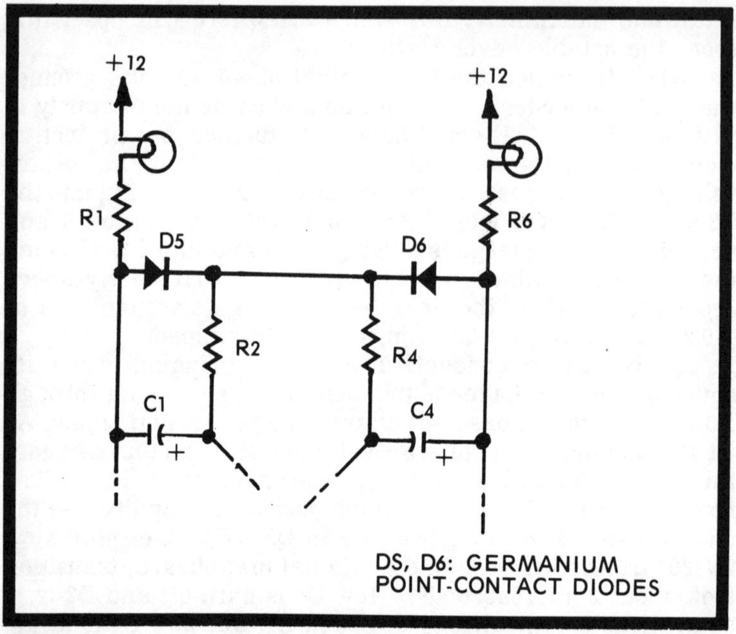

Fig. 5-7. The astable circuit is improved by adding diodes D5 and D6 so that it cannot fail to start when power is applied.

two more diodes as switches, and take transistor bias from the collector at higher voltage. Then, if both transistors are somehow momentarily saturated, they cannot find enough base current to stay saturated, and normal operation commences. Fig. 5-7 is only slightly different from Fig. 5-6. The added diodes are D5 and D6. This circuit runs a bit slower than its simpler predecessor, but starts reliably. It is capable of operation over a wide range of frequencies, and you can apply its principles to design new astables for many applications.

TRANSISTOR LOGIC CIRCUITS

Transistors make better logic circuits than diodes. They offer power gain, so that transistor logic circuits can be assembled into systems without using up the available input signal. And transistors invert the logic signal, since a logic swing one way at a transistor's base terminal causes a swing the other way at its collector. Often this inversion simplifies system design, but if it's inconvenient we simply add another transistor to turn the logic right-side up again. When we make our AND and OR gate circuits from transistors, we must indicate that the transistors invert the logic. We code this turnover into the circuit names by calling the transistor versions of our basic diode gates Not-AND and Not-OR, or simply "NAND" and "NOR."

The mathematics of NAND and NOR circuits is remarkably powerful. No other logic circuits are needed to assemble systems ranging from burglar alarm or heating system control right up to a computer's central processing unit, its memories and registers, and for the control of its input and output units. (The logical design would be clumsy, though.) NAND-NOR logic is the basis of modern logic system design. Fig. 5-8 shows a positive logic NOR gate (with extra logic flags for negative logic NAND gate operation). Its NOR gate operation is expressed as $A + B + C \quad \overline{X1}$, where the bar over the X1 signals the gate's logic inversion. The series silicon diode (type uncritical) drops the gate's output voltage so its down state is very near zero—it subtracts about 0.6 volts from the gate output voltage.

The positive-logic NAND gate in 5-9 is quite different from the preceding NOR gate. Its three transistors are in series; all

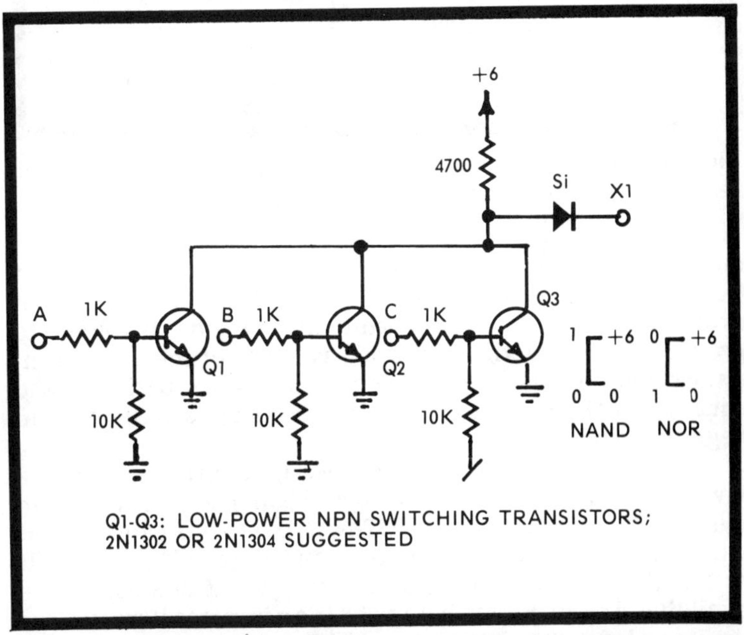

Fig. 5-8. The logical output of this NOR circuit is a zero if any input is a 1 in positive logic. In negative logic it is a NAND gate.

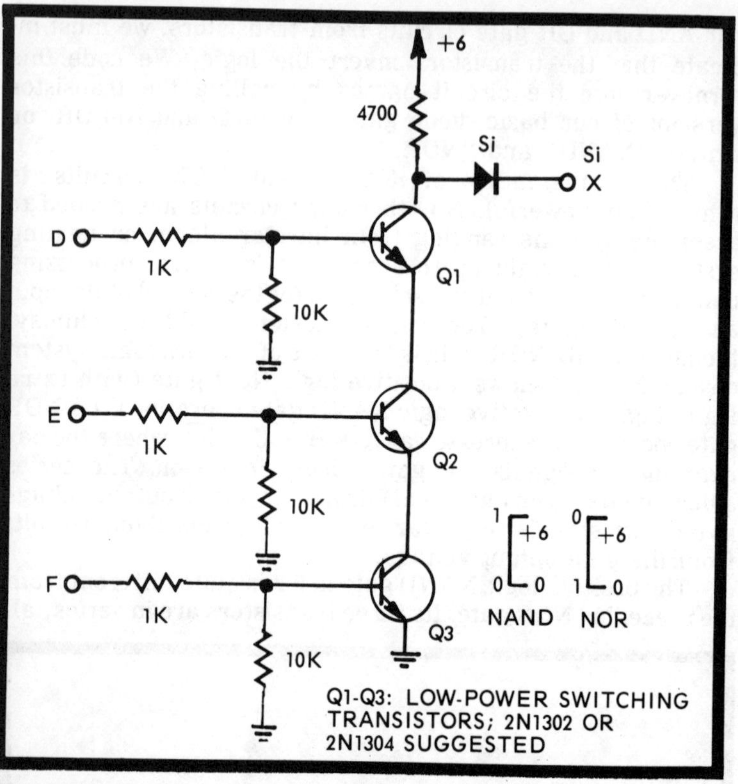

Fig. 5-9. A positive-logic NAND gate. Its output goes to zero only if inputs D and E and F are all 1s.

must be in conduction if the output terminal is to come down from supply potential. Here, the silicon diode's entire available voltage drop is needed to bring the down voltage to ground level, since the saturation voltages of the three transistors add up very near 0.6 volts. This gate's positive logic equation is $D \cdot E \cdot F = \overline{X2}$, and it becomes a NOR gate in negative logic.

An optional supplement to these gate circuits is in the inverter. This simply turns the logic signal over; a NAND gate followed by an inverter is a NOT-NAND or AND gate; a NOR gate evolves to a NOT-NOR gate or an OR gate. Either a NOR or a NAND gate can serve as an inverter, if nothing else is available. The NOR gate in Fig. 5-8 is an inverter if we use one input and leave the others open. And the NAND gate in Fig. 5-9 inverts if we tie two of its inputs to a fixed 1, probably the supply line, and feed a logical input to the remaining terminal.

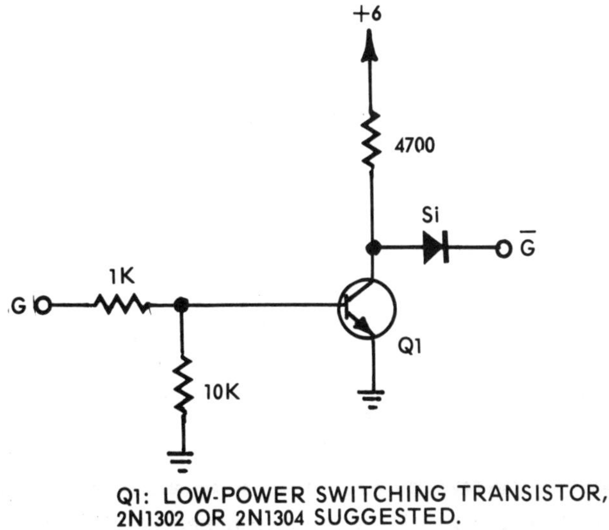

Fig. 5-10. A logical inverter or NOT circuit.

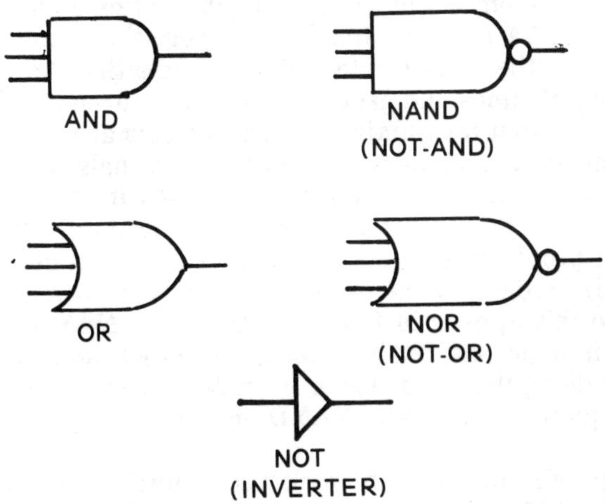

Fig. 5-11. Some logic symbols (many others are used). Each represents a complete circuit, although power supply connections are not indicated.

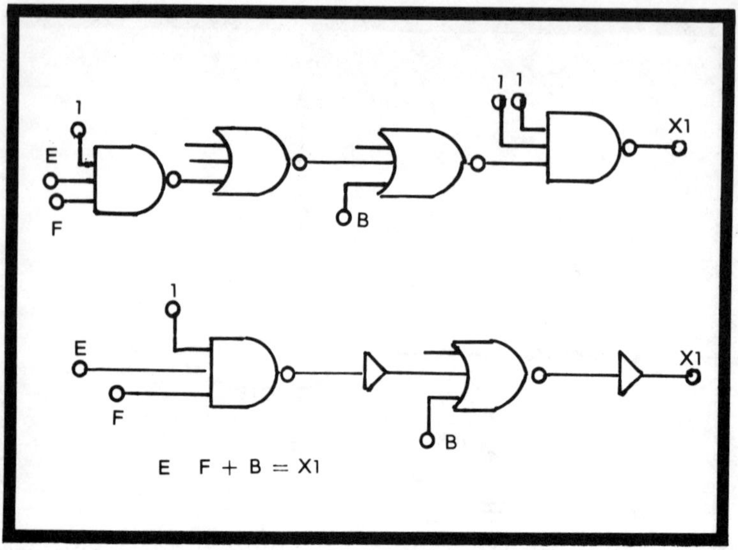

Fig. 5-12. A simple logic equation realized without and with inverters.

To save parts, it's nice to make up some inverters using a single transistor circuit like Fig. 5-10.

Logic system engineers avoid schematics with the detail of Figs. 5-8, 5-9, and 5-10. Instead, they use the symbols in Fig. 5-11. Notice the little circles added to identify the inverting gates. Supply lines and ground connections aren't shown. The input and output terminals of these symbols are marked with the same letters identifying similar terminals of the detail circuits to aid in associating the symbol with the circuit. In new design work, other labels, or often none at all, will appear.

Two problems suggest themselves for practice. How can we realize (real-ize) the equation E . F + B = X1? Two answers to this appear in Fig. 5-12. One uses NAND and NOR gates for logic and as inverters; the other adds inverters, thereby doing the job at lower cost. Now, can you realize the same equation using **only** NAND or **only** NOR gates, and inverters?

Logic circuits for lab work need simple real-world outputs. Fig. 5-13 shows two outputs easily made from commonly available parts. Transistor Q1 takes about 1 milliampere of base current to drive an inexpensive 500-ohm relay. Such relays are commonly available with SPDT contacts rated around 500 ma at 50 volts. Diode D1 absorbs the relay's in-

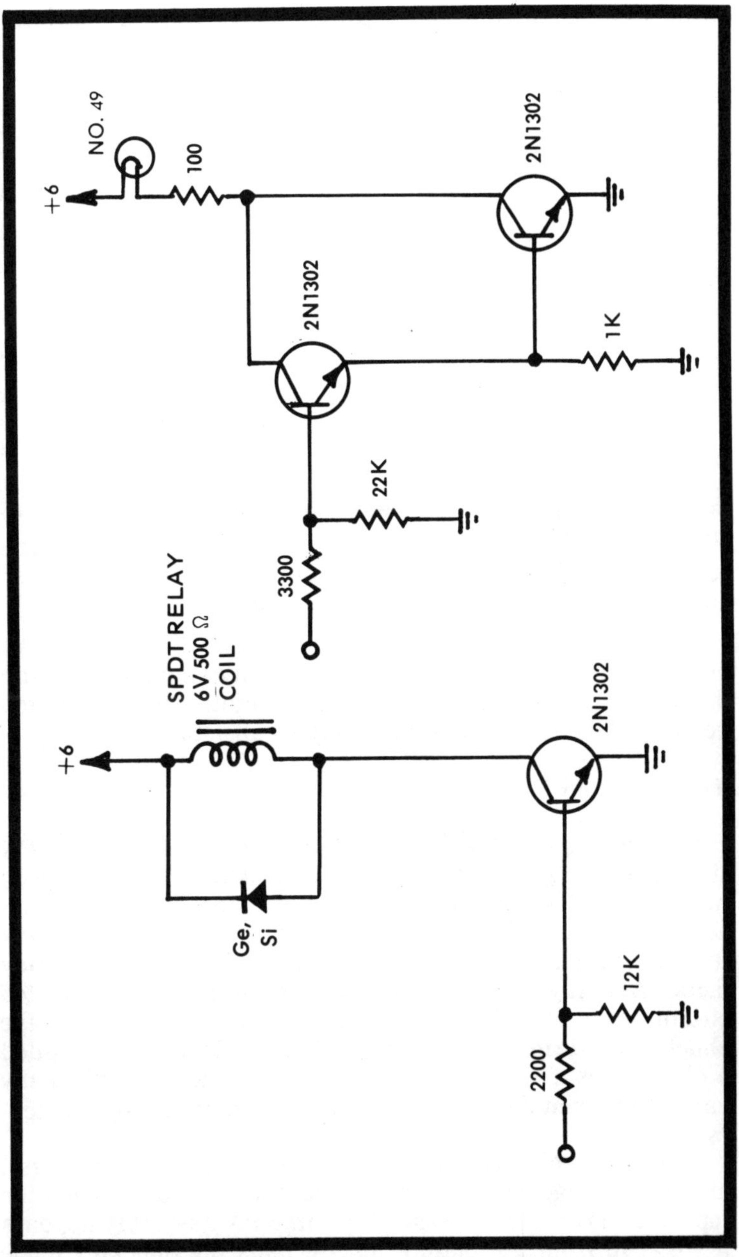

Fig. 5-13. Real-world outputs used with logic circuits. These are appropriate for control or indicator applications in real systems.

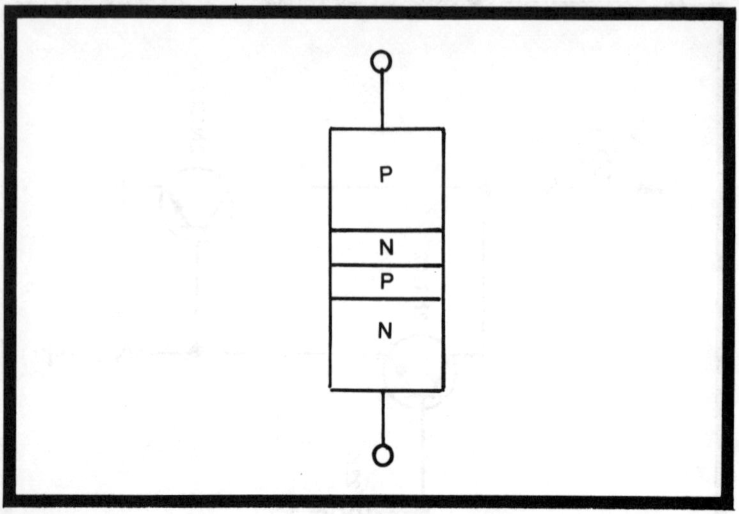

Fig. 5-14. A Shockley diode is a 4-layer PNPN device formed in a single piece of semiconductor crystal. At first sight it seems to be reverse-biased in either direction.

ductive kick, and the other output is simply an indicator lamp driven by a Darlington transistor circuit. The Darlington needs far less base drive than a single transistor.

THE SHOCKLEY DIODE

Elaborating a diode PN junction into a PNP or an NPN structure gives us a transistor. Transistor performance is more versatile than diode action, but evidence of diode ancestry remains. What happens if we add still another PN junction, making a PNPN device? We wind up with a Shockley diode. Actually, performance now resembles an integrated circuit rather than any diode or even a transistor. But the Shockley diode in its simplest form has two leads, so it's called a diode. It is a useful trigger and control device with many variations, including silicon-controlled rectifiers and switches.

The basic picture of a Shockley diode seems meaningless, at first. See Fig. 5-14. Then we notice if we read down from the top we discover a PNP assembly; an NPN assembly appears if we read from the bottom up. Perhaps the Shockley diode works something like a pair of transistors. Redrawing it to emphasize its inside transistor-like structure leads up to Fig. 5-15. This turns out to be a little-used but perfectly workable circuit called the "complementary bistable," a dual of the familiar parallel-transistor bistable. In operation, the two transistors in this circuit are either hard-on or hard-off. If Q1 is delivering no collector current, Q2 gets no base current and conducts no collector current so that Q1 gets no base current. Or if Q1 is hard-on it delivers a large base current to Q2, whose heavy collector current holds Q1 hard-on. A Shockley diode working like this really has two stable states, and evidently if it is off it will stay off.

A little transistor theory helps us see how the Shockley diode turns on. Earlier, we discussed transistor beta, the ratio of a transistor's collector-to-base current. In real transistors the conduction mechanisms are inefficient if they haven't much to do. Transistor beta falls off for very small collector currents. This roll-off is shown for some real transistors in Fig. 5-16, where we see that some work better than others, but at a low enough current beta always drops. When a Shockley diode is off, its leakage currents are well below that necessary to energize its inside transistors to betas near 1. It stays off because it isn't active enough to turn itself on.

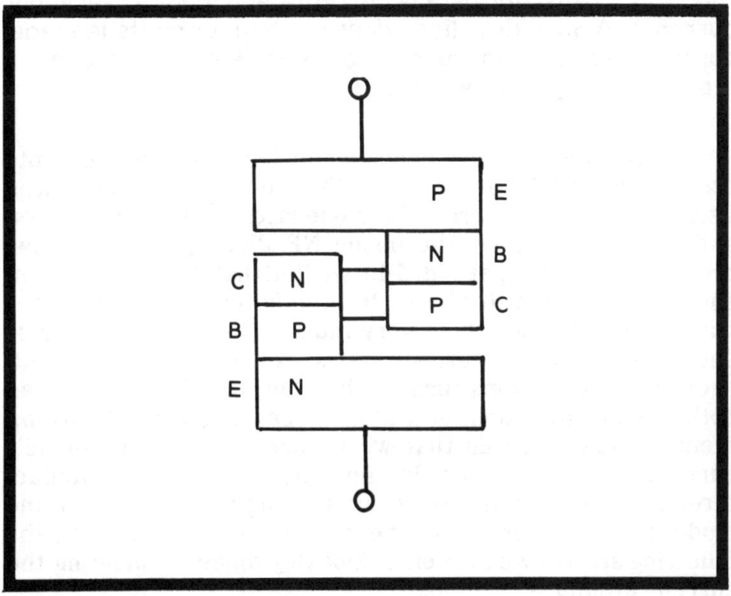

Fig. 5-15. Redrawing a Shockley diode makes it appear as a pair of conplementary transistors. This is very helpful in understanding how it works. Real examples of the complementary bistable can be assembled from silicon transistors.

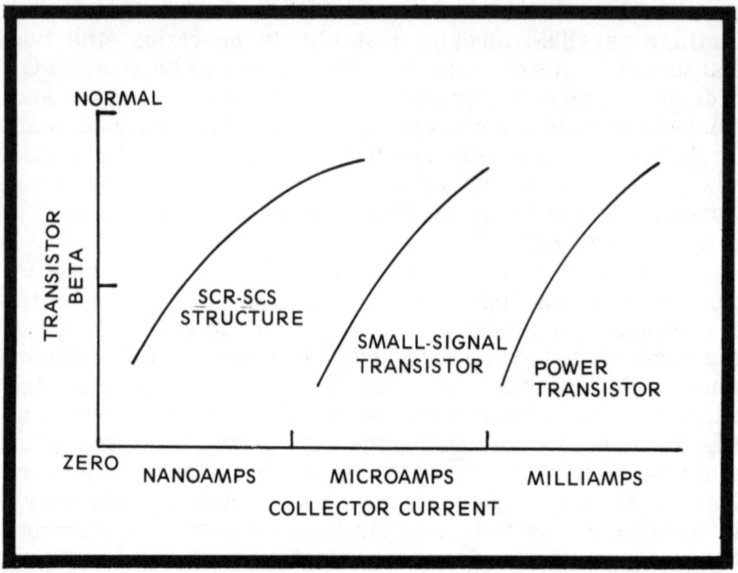

Fig. 5-16. Transistor beta is current-dependent at very low currents. A Shockley diode does not trigger on its leakage current because its current gain is well below one at typical leakage current levels.

We seem to turn on a Shockley diode by overvolting it into breakdown, but the process isn't that simple. The diode turns itself on. As voltage across the diode rises, leakage increases, and at a certain level the inside NP diode goes into zener breakdown. Looking at Fig. 5-15, we find that the zener current energizes the emitter-base terminals of both equivalent transistors, bringing them very much to life. Their betas rise and, when collector current starts exceeding base current, each transistor begins turning the other on. This stops when both are in saturation, and now we can think about turning them off again. To do that we reduce the Shockley diode's current to zero or to a value near it, listed in the manufacturer's specifications. In an alternating-current circuit the diode goes off every time the current goes to zero. In the following section we turn off a Shockley diode by shunting the current around it.

THE SCR AND SCS

With no terminals into its workings, a Shockley diode can be turned on only by overvolting it. This lacks finesse and limits circuit versatility. Shockley diodes with control terminals appear in two varieties: silicon controlled rectifiers (SCR), and silicon controlled switches (SCS). See Fig. 5-17. The "rectifiers" are fitted with a cathode gate terminal; the "switches" get an added anode gate terminal. Both can be turned on or off from either terminal; in real circuits it is much harder to turn off a PNPN device than to turn it on through these terminals.

SCRs and SCS serve in alarm circuits, where their simple once-triggered, stay-on characteristics simplify design and reduce costs. They are used as lamp drivers, where they work like saturating transistors but don't have to be held on by any outside circuits. They work in logic and memory circuits because, like bistables, they preserve a 0 or a 1, an off or an on, and are sometimes used in counters, registers, and even logic circuits. Since the power gain from control to output terminals is high, they can even be used in oscillators. They are used in industrial timing and comparison circuits where they convert some gradual change or difference into a sharply defined output. Large SCRs manage high powers, such as the 15-pound watercooled General Electric C500X1, which switches AC loads of 1200 amperes at 1800 volts, about 2 megawatts.

The following is concerned with SCRs, whose versatility extends far beyond the power supply applications implied by

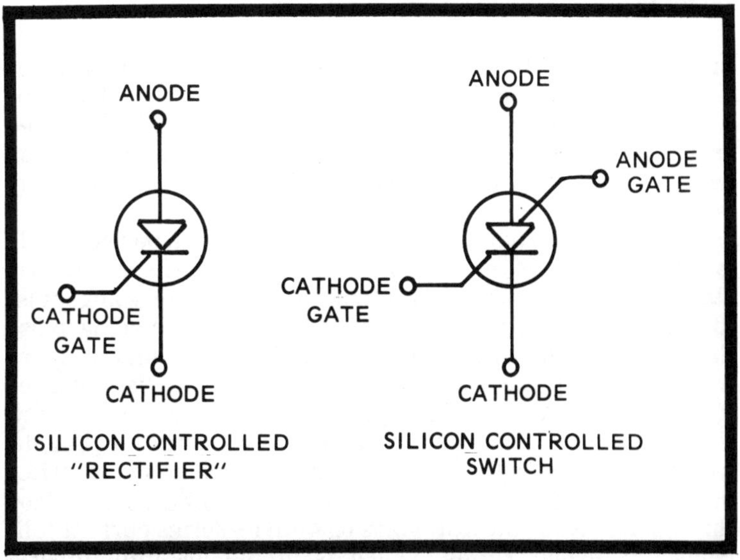

Fig. 5-17. The SCR and the SCS are about the same kind of PNPN device, but the SCS has an additional terminal.

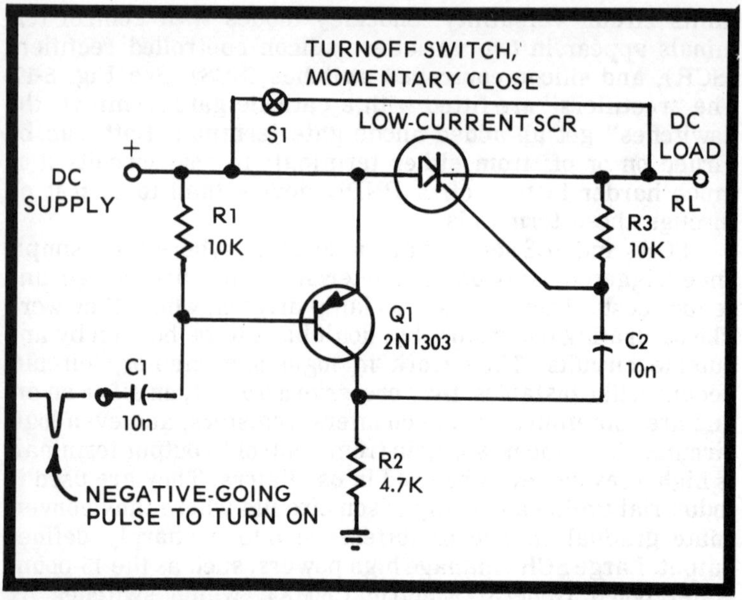

Fig. 5-18. A brief pulse will trigger this SCR unit on. In this circuit the SCR serves only as a switch. Closing S1 momentarily deprives it of operating current so that it turns off.

their "rectifier" label. Here, SCRs serve as power switches controllable remotely or by logic circuits. The first can be turned on electrically, but not off. And the second, presented as a separate circuit, provides a turnoff function. The two can be combined. We begin with the on-going control circuit of Fig. 5-18.

Initially, let's suppose the SCR is off, but the power supply is operating. This supply might be some batteries, for instance. The SCR will stay off until a negative-going pulse through C1 to the Q1 base results in a positive-going pulse through C2 to the SCR cathode gate. If a few volts of sharp positive-going pulse is already available, then Q1 isn't needed to manufacture it. The positive pulse forces a few microcoulombs of charge through the SCRs bottom PN diode, which invokes enough transistor and feedback action to switch the SCR on. Of course, there must be a closed current loop through RL. Now, the SCR stays on until its series current falls below its turnoff rating, which is in the low-milliamperes range. A very simple way to do this is close S1, which shorts the SCR briefly. With zero volts across it, the series current drops to zero, the conduction processes stop, and it goes off and stays off when S1 is re-opened.

A transistor circuit can shunt the current momentarily around the SCR just as well as a switch, and the circuit can be controlled from a remote site or through logic wiring. In Fig. 5-19 a silicon diode is added in series with the SCR to increase the voltage drop along the circuit to slightly over 1 volt. A power diode will do nicely here, although it probably isn't needed. When a positive pulse is applied to the Q2 base, this pulse appears as a stronger negative-going pulse at the Q3 base. This puts Q3 briefly into saturation. Now, Q3 has an emitter-collector voltage near 0.2 volts while in saturation, far less than the 1 volt or so across the SCR and its silicon diode. All the circuit current passes through Q3, which takes the place of S1 in Fig. 5-18. The SCR goes off and, as soon as the control pulse is exhausted, so does Q3. Since the above on circuit and this off circuit have nothing in common, they are easily combined as a circuit-controlled or remote-controlled switch that can manage large DC powers.

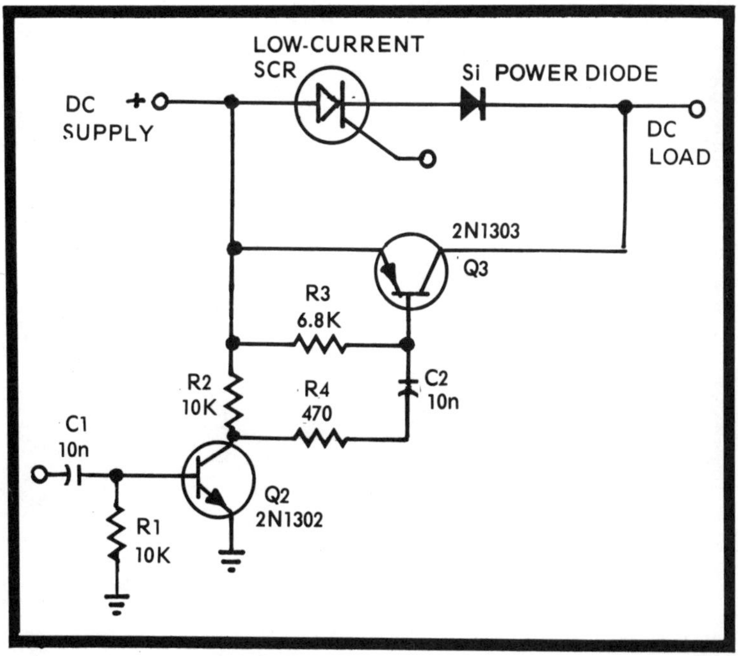

Fig. 5-19. This slightly more complex circuit turns the SCR off electrically. Basic action is like the switch of Fig. 5-18.

Notice that Q1, Q2 and Q3 are biased off rather softly by providing low-value resistors from the base to emitter terminals. This works fine at room temperature but may be inadequate in hot locations. Using silicon transistors in place of the suggested germanium types will reduce leakage. Smaller base-emitter resistors can be used, but this decreases power sensitivity. Perhaps some diodes can be fitted into these circuits to strengthen their resistance to collector-base leakage.

Chapter 6

Linear Transistors

If you have been following the book with circuit building and experiments, now you have some idea what transistors do. Transistors no longer look odd; you can design and assemble simple circuits with a fair idea what they will do and why they will do it. Your ideas about using diodes and transistors in indicator, logic and control circuits can be used to assemble many challenging and useful design projects. Perhaps you are thinking of sticking to switching circuits, but I recommend you have a try at the opposite side of electronics—"communications" electronics, the electronics of linear circuits. Just what a linear circuit is will be apparent soon; this type of circuit is no less important than switching circuits. A high-fidelity system, for instance, is a very good collection of linear circuits, or it should be. But the basic differences in circuits and emphasis don't mean you have to go out and find an entirely new set of components. The same transistors used in the preceding circuits will work here too—they are simply applied differently.

"LINEAR" OPERATION

Switching circuits ignore tiny signals, such as transients leaking in from nearby circuits, interesting electrical signals from distant stars, or communication from nearby radio transmitters. This is a virtue in switching circuits: it improves their reliability. But we are often interested in small signals and electrical events completely unlike the precise bang-bang on-off of switching systems. Maybe we like a hi-fi music; perhaps we are simply interested in effective voice communication with somebody three thousand or two hundred forty thousand miles away. There are hundreds of applications for what we call "linear" circuits.

A linear circuit is one whose output relates to its input in a gradual, smooth manner; an input of any reasonable size results in a recognizably similar output. When our linear circuits begin to show the least sign of switching circuit action we start talking about "distortion," "clipping," and worse.

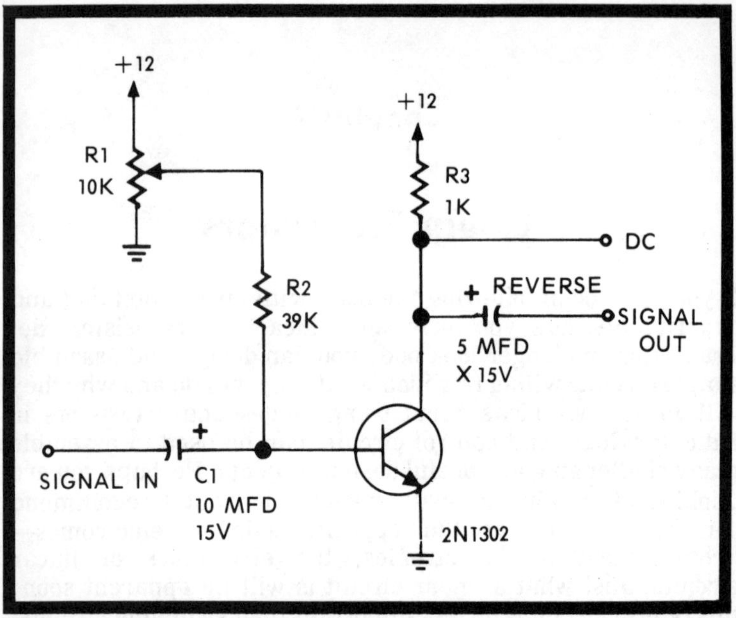

Fig. 6-1. A lab circuit designed to demonstrate transistor bias and input-output characteristics.

And we promptly set about correcting the bad component or design responsible for this malfunction.

Linear circuits work with about the same kinds of transistors as switching circuits. Marketing and catalog descriptions emphasize such differences, the designer's intentions make this distinction seem greater than it really is. Circuit design is far more important. A batch of good, inexpensive switching transistors turned up in the surplus market is probably an excellent purchase for amateur electronics work. The 2N1302 and 2N1303 germanium transistors peppered through this book are available new at little over 30¢ each because millions were made for computer construction.

We begin our examination of linear transistor circuits by observing a lab example. See Fig. 6-1. By adjusting R1 from ground to +12 volts, Q1 can be biased to any state from near cutoff to saturation. We estimate base current, which is very small, by checking the voltage at the upper end of R2 and then supposing its lower end is at +0.2 volts. This way we make good measurements using inexpensive test gear. Since the Q1 collector current much exceeds its base current, this terminal isn't upset by a multimeter's current drain.

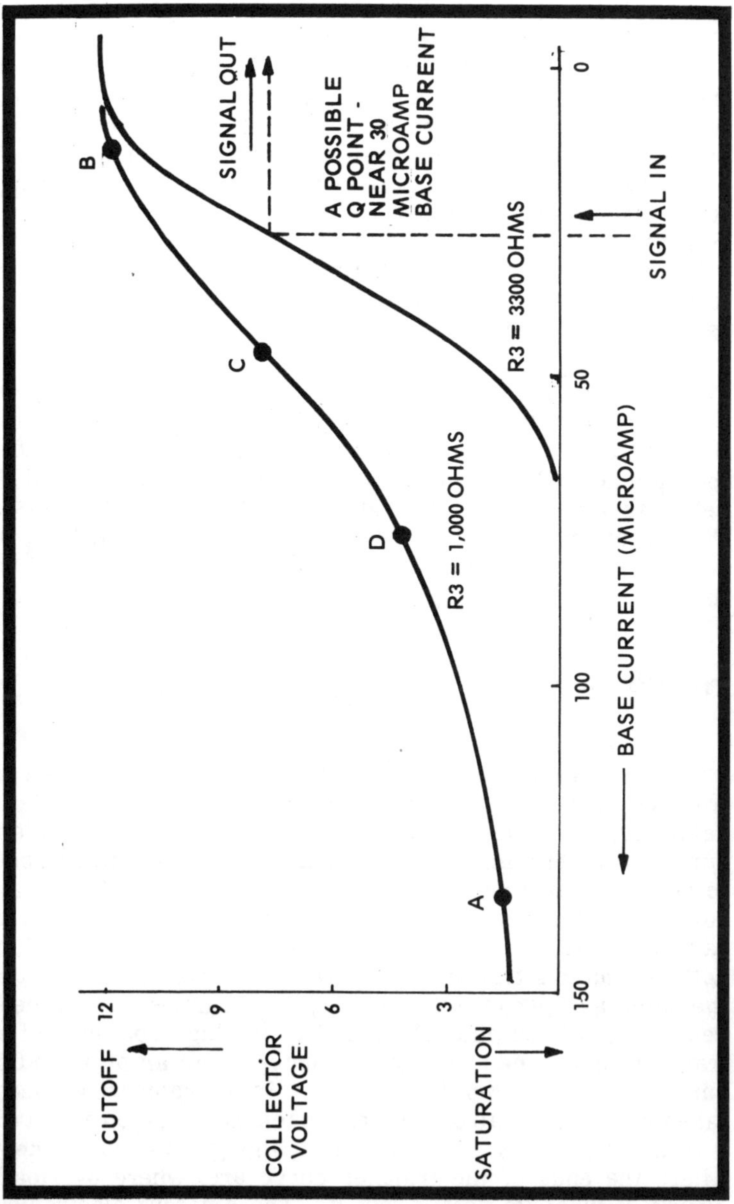

Fig. 6-2. Two DC input-output transfer curves demonstrating cutoff, saturation, and gain dependence upon load resistor value. A small audio signal can be applied to this circuit to compare the output with that predicted from the transfer curve.

There are two ways to make Fig. 6-2—smart and, to be nice about it, not-smart. The wrong way, we begin with the R1 slider at one end of R1 and advance it blindly in tiny steps, recording all results. Having collected lots of results we plot them all and then sketch in a line, discovering what came out. All this takes time and wastes every bit of it. A better approach energizes the brain first, as we think about what kind of results to expect. Since base current rises while advancing R1, collector current must go up too. This increases the voltage drop across R3; advancing R1 makes the collector voltage slide toward ground. With a meter on the Q1 collector terminal we swing R1 through its range and observe that this seems really to be the case.

Now we go in for detail. Near the R1 lower extreme we plot Point A; near the R1 other extreme, Point B. From our thinking of the preceding paragraph we could nearly finish the job right here. But, being careful, we now plot Points C and D before sketching in the finished curve. All this goes very fast. We repeat the test for a 3300-ohm resistor at R3, getting a considerably steeper curve. What does this steeper curve tell us?

Transfer Curve

The resulting graphs show us how some input condition is transferred to the output for all reasonable inputs. It is called, naturally, a "transfer curve." (All such plots are called curves; engineers grin when reporting, "This curve turned out to be a straight line..." but they say it anyway.) A similar curve could describe the performance of a fine hi-fi system; a radio receiver, or antenna gain as a function of its orientation with respect to a transmitter. "Transfer curve" means much the same as "performance curve," but carries the implied detail "input-to-output." We are often talking about part of a transfer curve when we say, for instance, an amplifier puts out 3.2 watts for a 6-millivolt input. Many engineering specs take this form, and all of these really tell us how steep a part of a transfer curve is. A detail is lost, though; we don't know where the ends of the transfer curve are, where overload occurs or what its details will be. The amateur electronics worker who thinks about transfer curves and their possible meanings develops a very useful mental tool. The meaning of the second transfer curve is it's an example showing us the greater circuit gain due to the steeper transfer curve.

Q Point

Let us suppose we choose a certain no-signal resting point for Q1's voltages and currents, in Fig. 6-2. At the moment this is the point where the zero-signal line is reflected off the transfer curve. In engineering terms, this is the "quiescent point," or simply the Q point. When we design a circuit we choose (or should choose) a Q point for each transistor. Then much of our work is concerned with circuits that will maintain the Q points we chose.

Many influences work against this, however. Transistor properties are temperature-dependent. A transistor's beta rises as temperature goes up; its beta may be 30 some chilly winter night outdoors and 70 the next summer in your sun-heated car. Its diode base-emitter voltage varies too, dropping roughly 2.5 millivolts per degree C temperature rise. Differences between an original transistor and its replacement may be considerable. And we must allow for aging components, batteries that run down, AC line voltage variation, and so on. Circuits that resist these influences aren't hard to design, but the matter deserves and needs close study by anyone who wants to design reliable circuits.

In designing circuits, we commonly choose a Q point by what looks like a guess. A running start on this might be, "Well, I think this is a 5 milliampere day." This sounds fantastic and is head and shoulders above ignoring the matter. Five milliamperes is a rather nice current for small-signal transistors. As a rough approximation, a transistor's Q point is chosen so that about half the available supply voltage appears across the transistor and so its collector power dissipation runs one-third to one-half maximum rating. Q point choice is less important with modern transistors than with older types, since transistor designers have been learning how to make transistor characteristics relatively immune to Q point changes. And if you actually breadboard the circuit, you get lots of chances to improve your thinking.

Signal-Bias Circuits

Where switching circuits give little thought to Q point details, beyond a moderate interest in how firmly transistors are off at high temperatures, linear circuits embody a complex relation between bias and signal wiring. That is, bias and signal connections are inseparably combined. Leaf through this book and examine the projects. Tracing connections will uncover two routes by which charge flows through each circuit. One is that followed by the steady DC currents that maintain quiescent conditions and power the circuit's working, and the other is that followed by what really concerns us—the signals. This brings up a pressing thought: What does the designer do when it looks like he needs two different resistances at the same point in the circuit? That rarely happens. The needs of one circuit or the other usually come out ahead. Where there is conflict the ingenious designer comes up with something, such as a large series capacitor that breaks a DC circuit but not a signal circuit, or a choke having a low reactance to DC and low frequencies, but which stops signal from leaking off into wrong parts of the overall circuit.

How To Design Circuits

The following advice is best viewed as a bit fantastic and perhaps highly visionary. The trained engineer uses lots of science for circuit design, so if you are getting by without his special training you have to come up with a substitute. Substitutes exist. A key point you may have missed is the engineer's design must work for all tolerances of all components; you can use cut-and-try to zero in on a Q point or adjust a detail of performance where he often cannot.

To design a circuit, or to understand it well, you might proceed as follows:

1. Think generally about the circuit, calling to mind what you know about all circuits of the general type. Try to see the circuit as already finished and yourself as a detective discovering the details, one by one.

2. As details emerge, add them to the schematic. Choose component values, circuit voltages and currents, interconnections that look about right. Remember this is all on paper. Then start working out further details with particular reference to bias voltages and currents. If these aren't right you cannot get good results; try concentrating on this part of the design problem for a little while.

3. Now attend to the signal side of it. Choose capacitors that are large enough, and no more. This reduces costs, and it also minimizes "tricks" that can develop in the circuit which didn't occur to you. Try to determine if the circuit will definitely do what you want it to do. When you can clearly see what should happen, you're making good progress.

4. Recycle. Look around until you've touched all bases firmly. Remember this isn't a baseball game. You can run the bases as many times as you like, and in a different order every time. It is important that when you are finished all your parts choices, voltages and currents should agree with each other. A well-chosen circuit can be detailed adequately for good results much easier than you probably believe. It just takes careful thinking, practice—and a certain amount of courage.

TRANSISTOR EQUIVALENT CIRCUITS

It's pretty clear what a circuit looks like to a transistor. It usually looks like a resistive load. If we cannot instantly see what this load's value is, an application of Thevenin's or Norton's theorems usually brings out details. And if that doesn't work, a ballpark guess will get things started right, at least. But, what does a transistor look like to a circuit?

An "equivalent circuit" illuminates details of a transistor's appearance to its surrounding circuits. A transistor equivalent circuit reflects the complexity of transistors, as compared to the battery versions we saw earlier. The transistor equivalent circuit contains more parts. Components shown connected to the input terminal account for this terminal's appearance to incoming signals. And a Thevenin or Norton dependent generator supplies facts concerning the transistor's output properties. "Dependent generator" simply means that the equivalent voltage or current, instead of being fixed as for a battery, now depends upon the signal applied to the input terminal. A little note beside it describes what the relation is.

Part of the highly impressive appearance of some engineering texts is due to their elaborate and (to the untrained eye) apparently meaningless equivalent circuits. The general effect is multiplied in some books by a generous peppering of conversion tables for relating key points of different equivalent circuits. The overall result almost seems to have no connection with electronics. Don't you believe it. But we will stick to equivalent circuits related closely to matters we've already discussed, or which are nearly self-evident. But first we must put in a little "gound" work.

Signal Ground

In historical times and up until 1950 or so, technicians checking circuits hooked one test lead to the metal chassis carrying the electronics and referred all voltage and signal

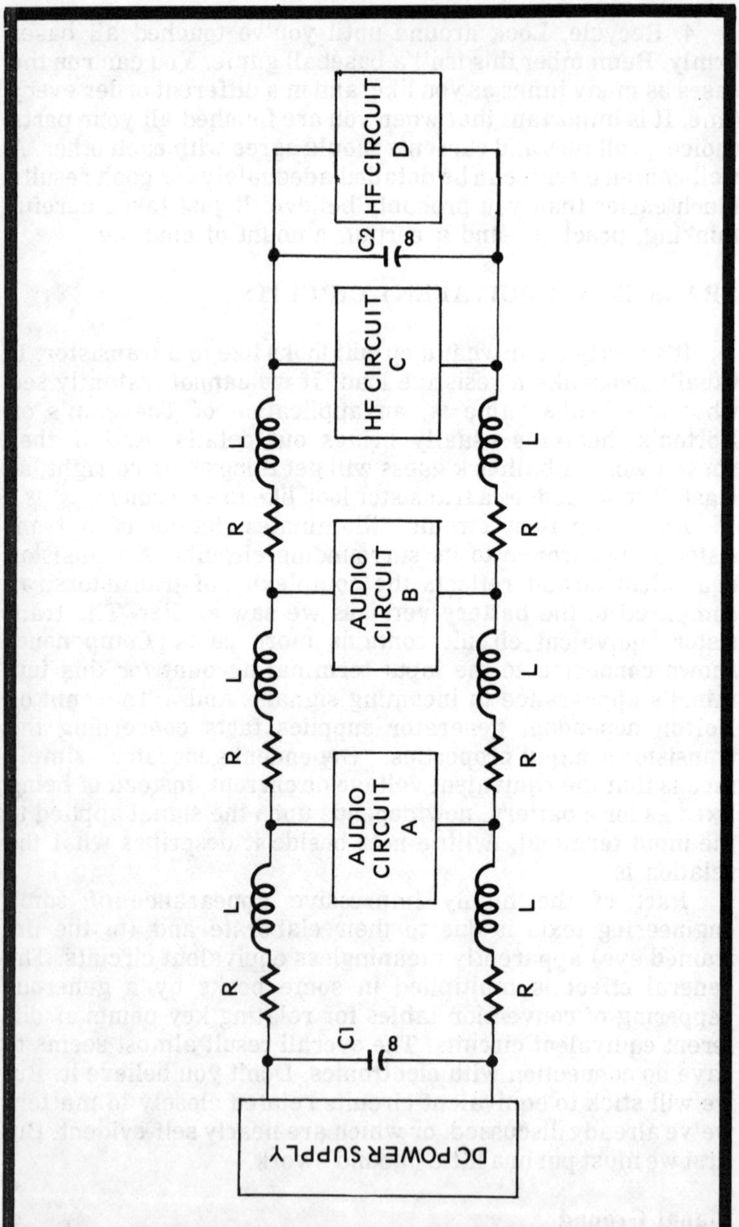

Fig. 6-3. An easily missed important detail of circuit operation is the electrical characteristics of its supply leads.

tests to that terminal. They called it "ground," a term held over from earlier days. Way back then, a true earth connection was both far easier to achieve (smaller buildings, simpler circuits, etc.) and thought to be more important than now. "Ground" simply translates, in that thinking, to "earth connection." These days we hardly ever think about an earth connection, except for safety purposes.

To further emphasize how artificial the "ground" concept is, think about a ship in space, a man orbiting in a spacesuit, or a small communications satellite. This limited, isolated environment upsets the circuits not at all, so evidently electronic circuits see "ground" in their own special way. An important part of your developing picture of how things work will depend upon your learning to discover circuits' perspective on "grounding."

Since supply and bias voltages and currents all refer to the power supply leads, and all signal voltages and currents are derived from these flows, we now know where our circuit thinks ground is. Our circuit **knows** ground is the supply leads—all of them. The designer who misses this highly important point is sure to get into trouble. In Fig. 6-3 we have a general kind of electronic system consisting of two audio circuits and two high-frequency circuits. It happens that this time there are only two supply lines. Each circuit has its own signal voltages and currents, and we see that these cannot refer to anything except the supply lines. Looking at audio circuit B we find it is impressing signal current into the supply line, and that this current flows through parasitic resistances and inductances R and L, around through C1, and returns to B's other terminal. The R and L parasitic components, representing power wiring resistance and inductance, are very small, but **they are not zero.**

This means B's outside signal current develops a small voltage across A's terminals, no matter how large we make C1. Evidently, a large C1 helps, but it isn't a complete remedy to the problem of a signal current generating voltage across A's terminals. But at audio frequencies there will be little feedback because the R and L parasitics are extremely small. The value of C1 usually interests the designer more, or it should. If a circuit is troubled by motorboating or other unwanted oscillations, an early step in curing this is to look for signal voltage on the supply lines. Surprisingly often, some turns up there and decoupling it somehow cures apparently tough problems easily.

Now we look at HF circuit D. "HF" means somewhere around 200 kHz or 2 MHz up. This circuit impresses signal

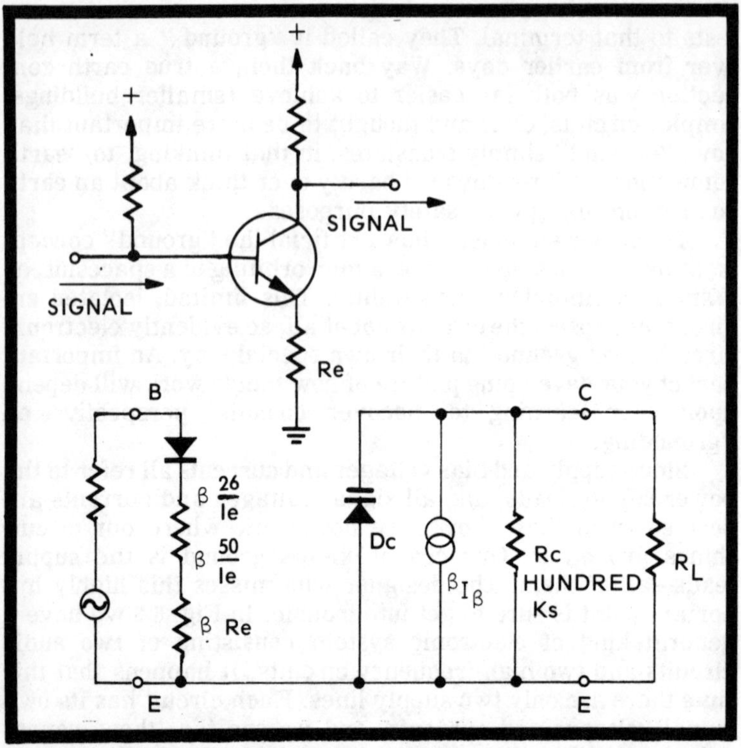

Fig. 6-4. A simple common-emitter transistor equivalent circuit. This is too simple for real engineering, but it gives good results when some cut-and-try is acceptable.

currents on the supply lines too, but at high frequencies the parasitic R and L are enough to isolate D from the signal—decoupling influence of C1. As a result appreciable signal appears across C's supply terminals, leading to regenerative—or degenerative—feedback. The remedy is simple. We just add capacitor C2 to bypass signal appearing in this part of the supply lines. C2, like C1, should have a reactance of only a few ohms at signal frequencies, and now short leads are very important.

Equivalent Circuits

Our three equivalent circuit types reflect the three transistor terminals available to be fixed at signal ground. This gives us common-emitter, common-base, and common-collector equivalent circuits. Common-emitter applications outnumber all others, and are a good general catchall when coming up with new ideas. But as circuits go, the designer who thinks too much in common-emitter terms is short-changing himself.

An important detail in the common-emitter circuit is that it's the only one that inverts signal polarity. The common-base and common-collector circuits develop same-polarity outputs; when the input signal is swinging positive so is the output signal. This is why the common-emitter dependent generator equation is preceded by a minus sign, but the others are not. Each of the following appears with a Thevenin generator signal source and a resistive load, RL. A Norton equivalent is an equally plausible candidate for supplying input signal, and RL can be replaced by the input terminal equivalent components of another transistor equivalent circuit. Diodes Db and De are supplied to remind us not to overdrive the common-emitter and common-base circuits, but since the com-

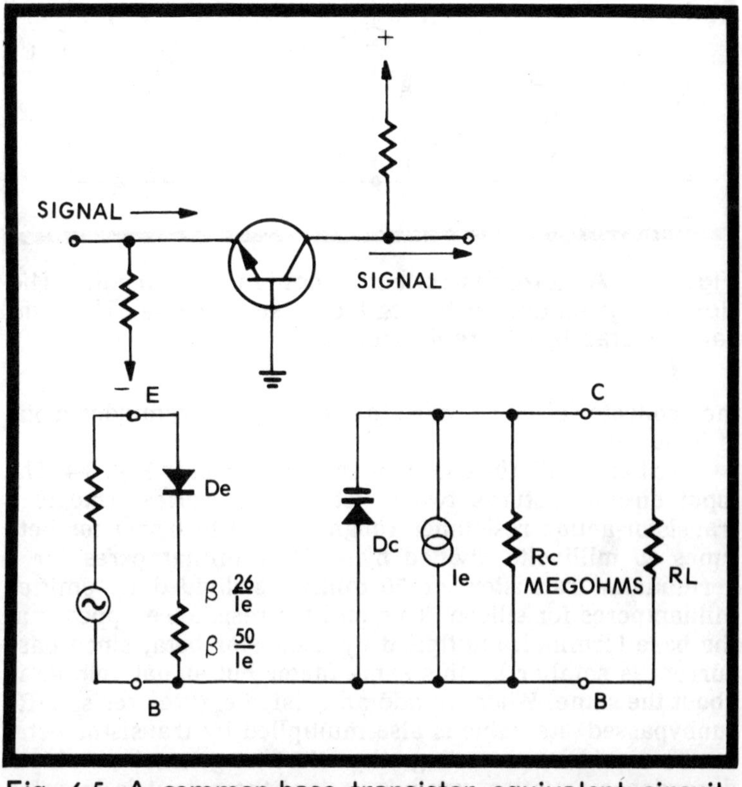

Fig. 6-5. A common-base transistor equivalent circuit. Notice the output dependent generator develops the same current that enters the emitter terminal.

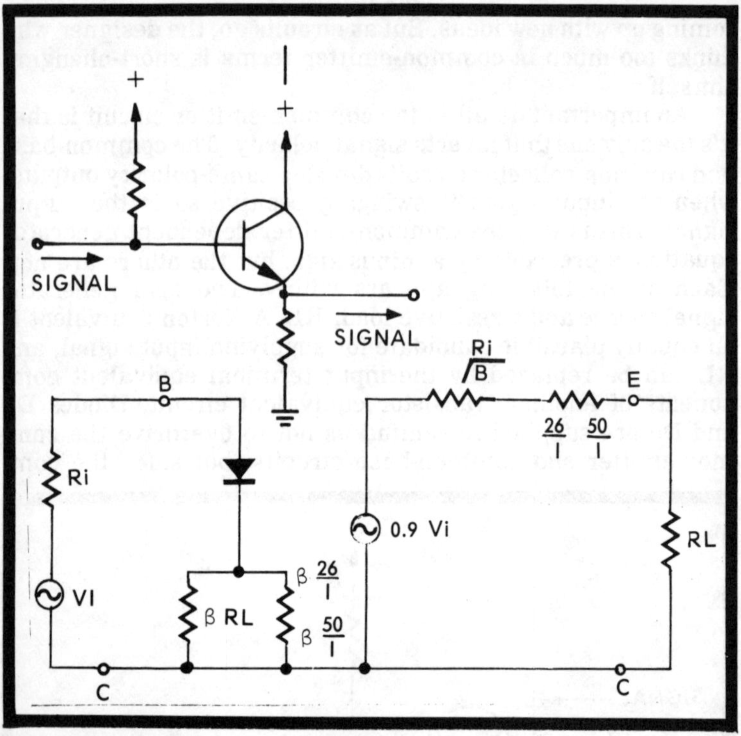

Fig. 6-6. A common-collector equivalent circuit. The signal is grounded to the collector supply line. This can cause instability in real circuits.

mon-collector circuit is hard to overdrive, no reminder diode is included.

We begin with the common-emitter circuit in Fig. 6-4. The input circuit features two resistances in series. One is a transistor-action resistance roughly equal to transistor beta times 26 millivolts divided by emitter milliamperes for a germanium transistor, or 50 millivolts divided by emitter milliamperes for silicon. This emitter resistance appears at the base terminal, multiplied by transistor beta, since base current is smaller by this same factor but signal voltage is about the same. When we add an outside emitter resistor Re (unbypassed) its value is also multiplied by transistor beta.

The minus preceding the Norton generator's input dependence equation reminds us that the output is inverted with respect to the input. A reverse-biased diode capacitor supplies a collector capacitance that depends on the Q point and which must be small for high-frequency work. It also reminds us we can borrow diode capacitors from our signal transistor box if we want adventure or reduced parts cost. Then the dependent generator goes to zero (no base input) and if we're lucky our diode capacitor works fine. If this sounds dubious, be sure to try it.

Our common-base equivalent circuit has to get by without the beta times resistance multiplication of the common-emitter circuit since the input goes right into the emitter diode terminal. See Fig. 6-5. If we are designing our own circuits, that's no handicap. Maybe we will choose a common-base circuit because we want a very low input resistance. But now our output equivalent generator puts out the same value current as our input signal, which looks bad at first. The circuit's power gain is represented by our transition from a low-impedance circuit (10-50 ohms typical) to a high-impedance output.

The common-collector circuit looks upside down because, if we are to follow the pattern of the other two circuits, we must remind ourselves the collector voltage supply line is now the signal ground. See Fig. 6-6. This particular detail is why we opened with a discussion of "Where is the signal ground?" A simple circuit like that in Fig. 3-36 contains a common-emitter amplifier followed by a common-collector amplifier to reduce the output resistance. The input stage sees its minus supply as "ground" and the output the plus supply as "ground"; this shows us that both can be equally important.

The common-collector circuit gives us a relatively high input resistance, but this is composed of two resistances now appearing in parallel. One is the beta times the basic emitter resistance, and the other is a resistance borrowed from the load—beta times RL. Bias network resistance usually comes in here, too, in real circuits. This triple influence rather degrades the high input resistance that seems available from the emitter follower. But good design work can overcome these loadings enough to obtain input resistances in the megohm range. Notice a slight input-to-output voltage drop and low output resistance indicated by the use of a Thevenin equivalent here. Also, the input source's Thevenin equivalent turns up in the output **divided** by transistor beta. It is added to the emitter diode resistance to make up the emitter follower's output resistance.

Very important: The preceding equivalent circuits are all presented in NPN types to avoid drawing everything twice. This is no vote for NPN devices solely. The PNPs are equally important, or more so if you cannot work as easily with one as

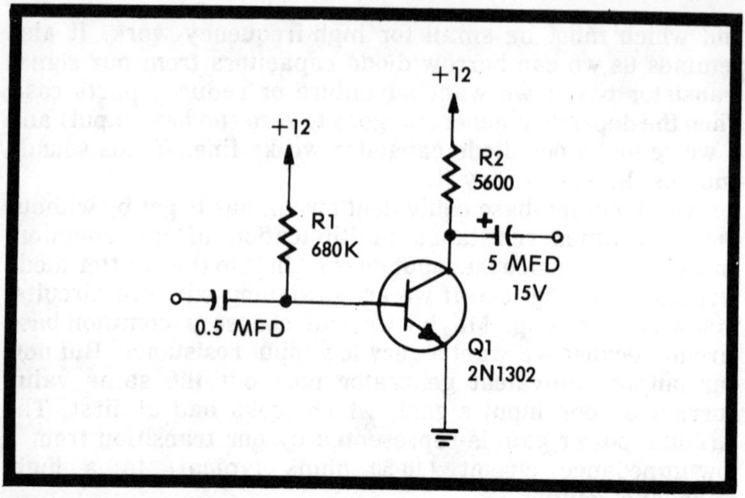

Fig. 6-7. This circuit features extreme simplicity and very poor bias stability. It's usable at room temperature for small-signal amplification.

the other. For maximum benefit from these equivalent circuits, practice drawing them and thinking about them until you can work equally well with either way of thinking.

BIASING

The following practical examples are actually meldings of two circuits each—a signal circuit and a bias circuit. For maximum benefit from reading this section, trace these two kinds of interconnections and work at seeing what each part does. Build lab models (breadboard style), observe their performance and your own, and try your hand at designing variations. This practice serves the same purpose as repetitively throwing a baseball, tossing baskets, or other sports training. You get a feel for how things go together and interact, and gradually real schematics will take on important new meanings you never noticed before—even if you've been working years over them.

Cut-And-Try Bias

Our earlier transistor experimental circuit appears pared for real work in Fig. 6-7. It isn't a very good circuit, since it is too sensitive to temperature changes and to device replacement. On the other hand, it uses a minimum number of parts. Choose an R1 for some appropriate Q point and then plug in some new transistors. You'll find considerable Q point variation. Yet this circuit does turn up in some commercial battery radios.

Feedback Bias

If we let the transistor take a hand in choosing its own Q point, our cut-and-try bias works better (Fig. 6-8). When the transistor is hard-on, its low collector voltage won't permit enough bias to keep it there; the transistor cannot go hard-off either. A fair Q point is a distinct possibility and with a little math you'll do well easily. The simplest version of this circuit feeds back signal from collector to base, but a bypass capacitor stops this.

We can improve this system (Fig. 6-9) and gain a control point if we return the base bias to a turnoff-polarity voltage. This arrangement appears in the sample schematic of Fig. 3-36. Returning the lower base resistor to a negative supply increases its appropriate value, giving the transistor strong

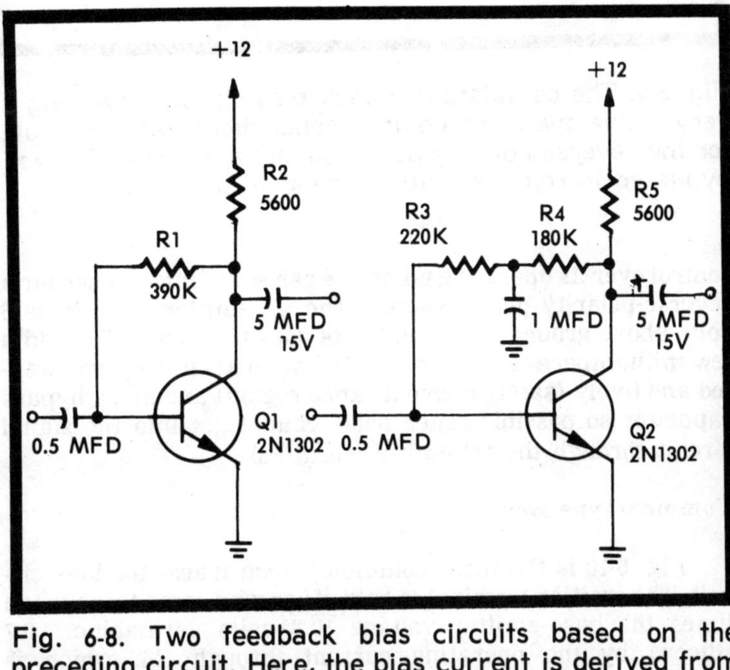

Fig. 6-8. Two feedback bias circuits based on the preceding circuit. Here, the bias current is derived from the collector terminal. The transistor takes a part in fixing its own operating point, which improves stability.

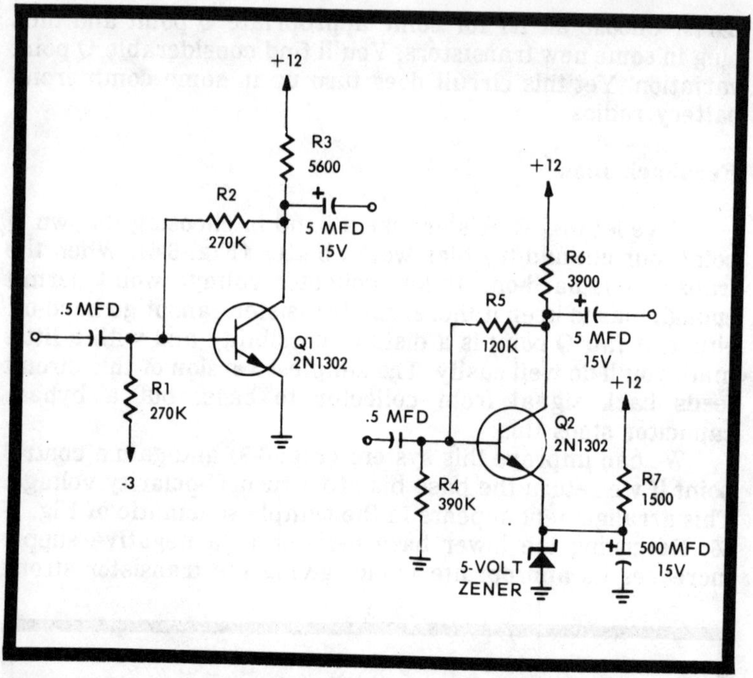

Fig. 6-9. The complete feedback bias circuit, featuring a very stable operating point. A zener diode can substitute for the reverse-polarity power supply and should be well bypassed to remove natural zener noise.

control over its operating point. We can substitute a zener for a reverse-polarity power supply. The Q2 emitter is perhaps 6 volts above ground, so it thinks R4 goes to -6 volts. We add a few milliamperes from the supply line to keep the zener well-fed and lively (safely above its knee region) and add a bypass capacitor so possible zener noise cannot get into the signal circuit through the Q2 emitter terminal.

Common-Type Bias

Fig. 6-10 is the most commonly seen transistor bias circuit. The emitter terminal is held off ground some three to ten times the base-emitter voltage (0.2 volts germanium; 0.7 silicon) by the operating current through R4, which is bypassed for signal frequencies. This capacitor sees a Thevenin generator, which is the Q1 emitter terminal, just like

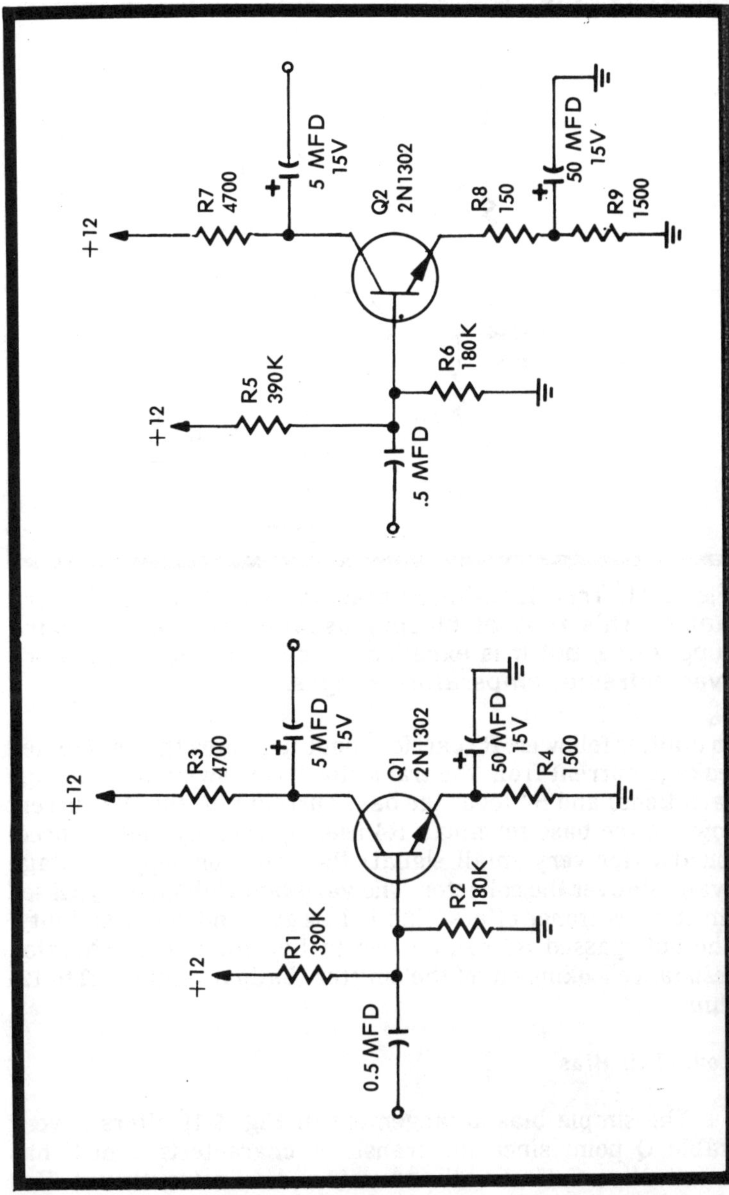

Fig. 6-10. Conventional biasing circuits. A small unbypassed resistor in the transistor's emitter circuit will improve linearity. The voltage across the entire emitter resistor is roughly two to ten times the transistor's emitter-base voltage.

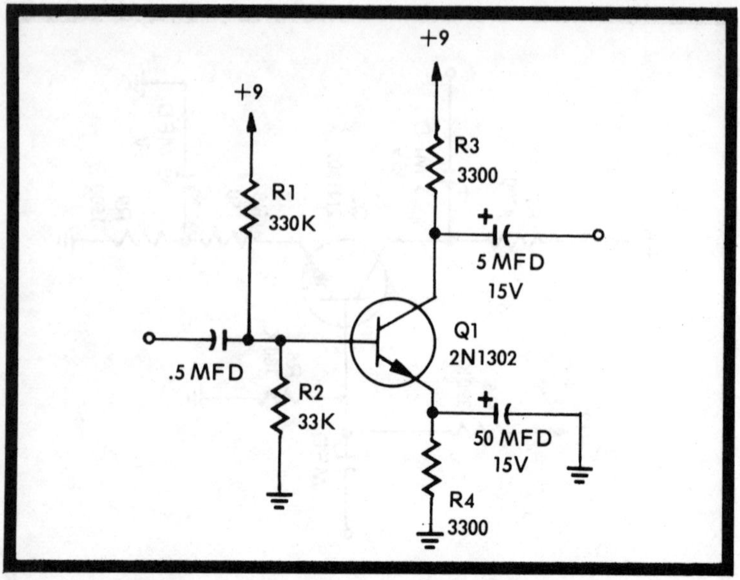

Fig. 6-11. True long-tailed biasing circuit, which is very stable. This type of biasing usually requires an extra supply line, but it is excellent in circuits that must work over extreme temperature ranges.

an emitter follower. R2 carries about ten times the anticipated leakage current from the transistor base (swamping leakage variations) and R1 feeds the base current plus this R2 current down to the base terminal. R4 uses up roughly half to three-fourths (for very small signal) the collector supply voltage available over the collector. The variation with an unbypassed emitter resistance offers slightly less gain and better stability. The unbypassed R8 can be two to five times the estimated resistance looking out of the emitter (ballpark figure, 22 to 120 ohms).

Long-Tail Bias

The simple bias arrangement in Fig. 6-11 offers a very stable Q point since the transistor characteristic drift has practically no effect on the emitter and collector current. You can easily design a long-tail bias circuit that works at extremely low temperatures, or accepts silicon and germanium small-signal transistors interchangeably. Q has no control over its operating point, which is determined by R4 and to a lesser degree by R1 and R2.

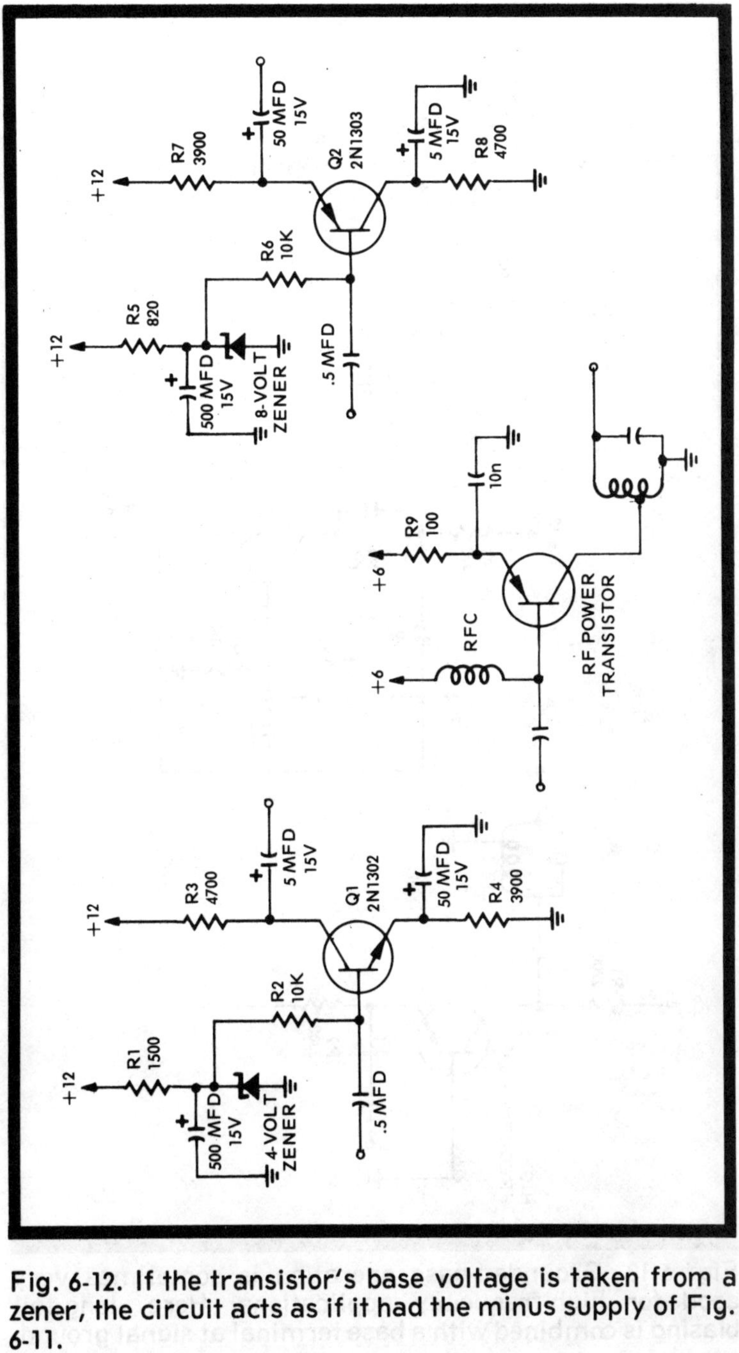

Fig. 6-12. If the transistor's base voltage is taken from a zener, the circuit acts as if it had the minus supply of Fig. 6-11.

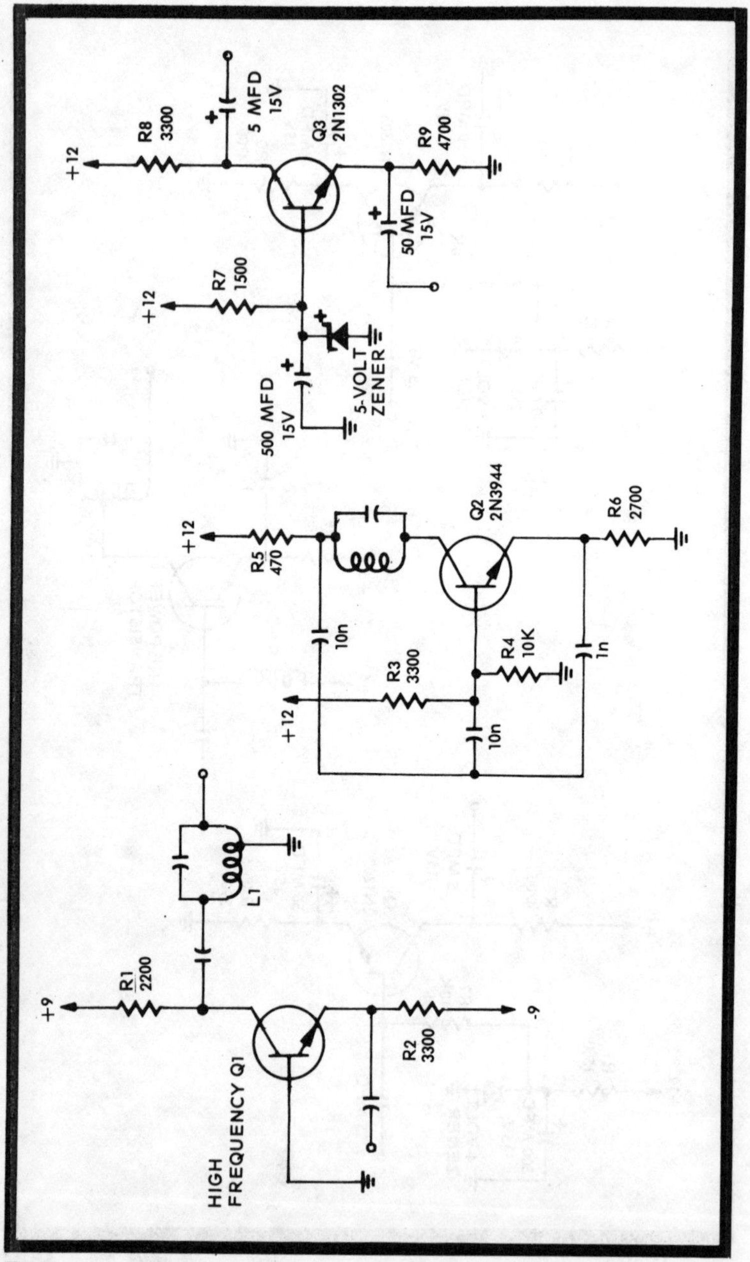

Fig. 6-13. Grounded-base operation is sometimes very good for high-frequency applications. Here, long-tail biasing is combined with a base terminal at signal ground.

Common-Emitter Signal Circuits

In Fig. 6-12 we get some ideas about zeners in biasing. Here a zener, rather than a divider network, provides a base reference voltage. This enables us to fix the operating point firmly, or to use long-tail biasing without adding an extra power supply. In the next example we find how to use a PNP transistor with a positive supply, sometimes a very handy arrangement.

An example is the PNP transistor frequency multiplier that works with its collector at DC ground. This can be mechanically useful at high frequencies. The base-emitter diode is shorted at DC through a small RF choke, so that is conducts only on half cycles. The resulting distortion, very bad at audio, makes this an effective frequency multiplier. If the emitter resistor is fairly large, the transistor current is sharply regulated, preventing transistor damage by over-drive.

Common-Base Signal Circuits

The real possibilities for adventure in circuit design begin to appear when we get away from the very common common-emitter basic circuit. If we choose a common-base approach, now we have a choice of connecting the base directly to a ground plane for RF work, or to a voltage divider or zener diode for other applications. It depends on how many power supplies we want to use, plus other details.

In Fig. 6-13 we begin with a very simple RF amplifier. This circuit needs good input loading, or it becomes a collector-to-emitter feedback oscillator. Long-tail biasing makes this very stable. A grounded tap on the output tuned-circuit coil (L1) varies the output coupling; about a 4:1 turns ratio gives a 16:1 impedance matching, where the source impedance is almost perfectly equal to R1. This tapped-inductor circuit, if you are watching for such details, is a dual of the commonly used pi-section tuner. This variation gets by with fewer parts.

Next we find a zener-biased long-tail circuit using a single power supply. The Q3 base terminal is held at signal ground by the same capacitor that is bypassing possible zener noise; it may be necessary to parallel an RF capacitor with a larger electrolytic of a few microfarads. Finally, there is an isolated-oscillator circuit. It is long-tail biased for stability, but its signal current loops are completely isolated from the supply lines. Its output is link-coupled to where needed.

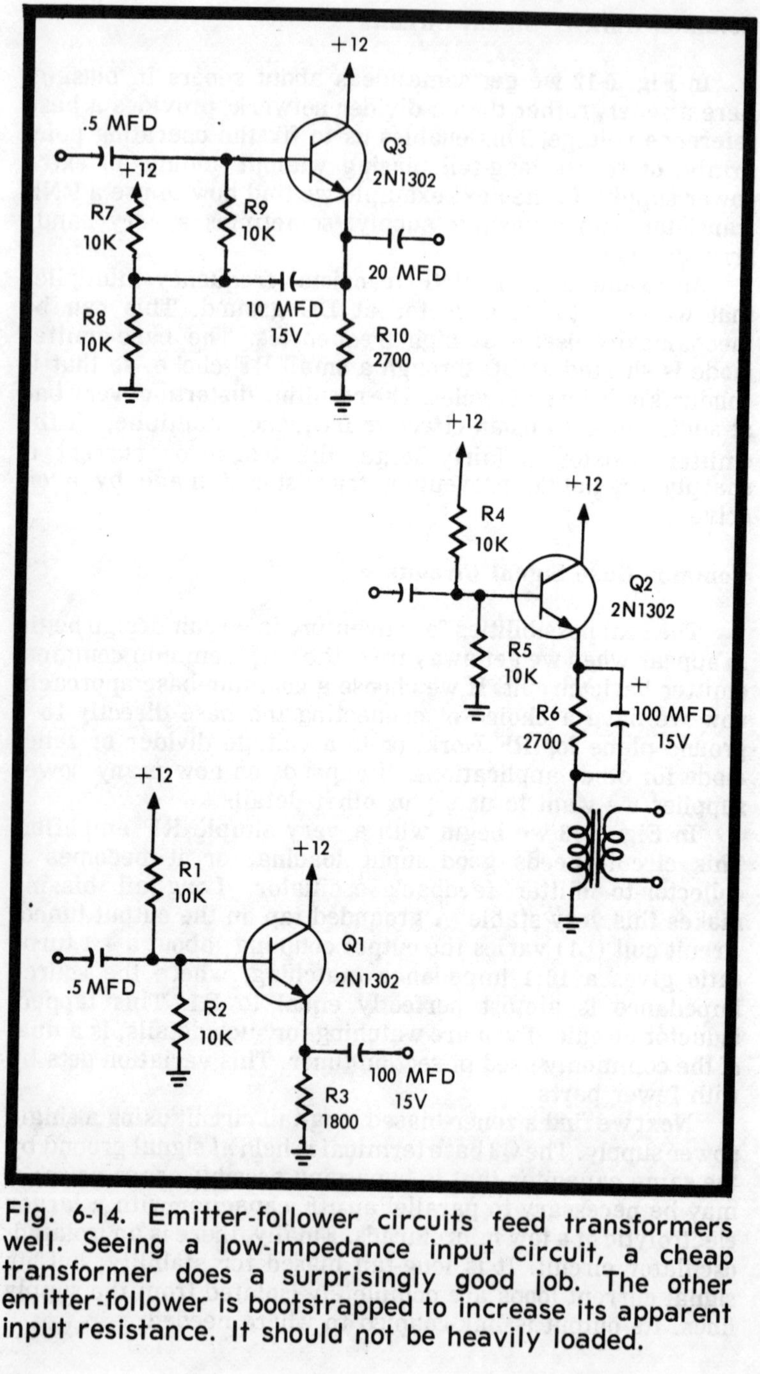

Fig. 6-14. Emitter-follower circuits feed transformers well. Seeing a low-impedance input circuit, a cheap transformer does a surprisingly good job. The other emitter-follower is bootstrapped to increase its apparent input resistance. It should not be heavily loaded.

Common-Collector Signal Circuits

In Fig. 6-14 are three variations on the circuit frequently called the "emitter follower." This circuit's style of operation keeps its emitter voltage near its base voltage, seeming to follow it as a constant difference of 0.2 volts (germanium) or 0.7 volts (silicon). This circuit is useful for getting a low-impedance signal from a high-impedance source.

The simplest variation on the emitter follower simply reduces signal impedance. Q1 is conventionally biased, but has no collector resistor. It needs none, with its signal appearing at the emitter terminal. The output impedance of this circuit can be as low as a few ohms. If we let the emitter follower feed a transformer, we get a still lower output impedance, yet the circuit remains simple. We must now bypass R6 so that audio signal is fed directly to the transformer winding. Transformers are very happy in low-impedance circuits like this, which means a cheap one will work well. A 6-volt filament transformer, not intended for audio work at all, will drive low-impedance phones very well. If you don't think so, try it. Frequency response will be flat from 20 Hz to 200 kHz or better.

Finally, we come to the bootstrapped emitter follower. Here, some of the signal appearing at the emitter is applied to the base bias circuit. Suppose R9 were 1000 ohms, and nine tenths of the input signal appeared at its lower end. As much signal current flows into R9 as would flow into a ten times greater resistor going to ground, so that R9 looks like 10,000 ohms. Real bootstrapped emitter-follower circuits like this one can develop input impedance in the megohm range.

"WHISTLERS"

Way back in 1886 an Austrian scientist studying electric circuits heard curious descending tones in some of his equipment. His work was widely ignored, but turned up again when Barkhausen, 32 years later, was finding ways to tap Allied telephone lines during WW I. Since then, whistlers and related low-frequency electrical noises have been studied all over the world, using antennas of all sizes up to one that was strung across the Grand Canyon. Today, we know what whistlers are.

They are one kind of electrical noise resulting from natural events in and near our atmosphere. Whistlers are lightning static, after it's traveled far into space and returned along a route apparently corresponding to magnetic lines of

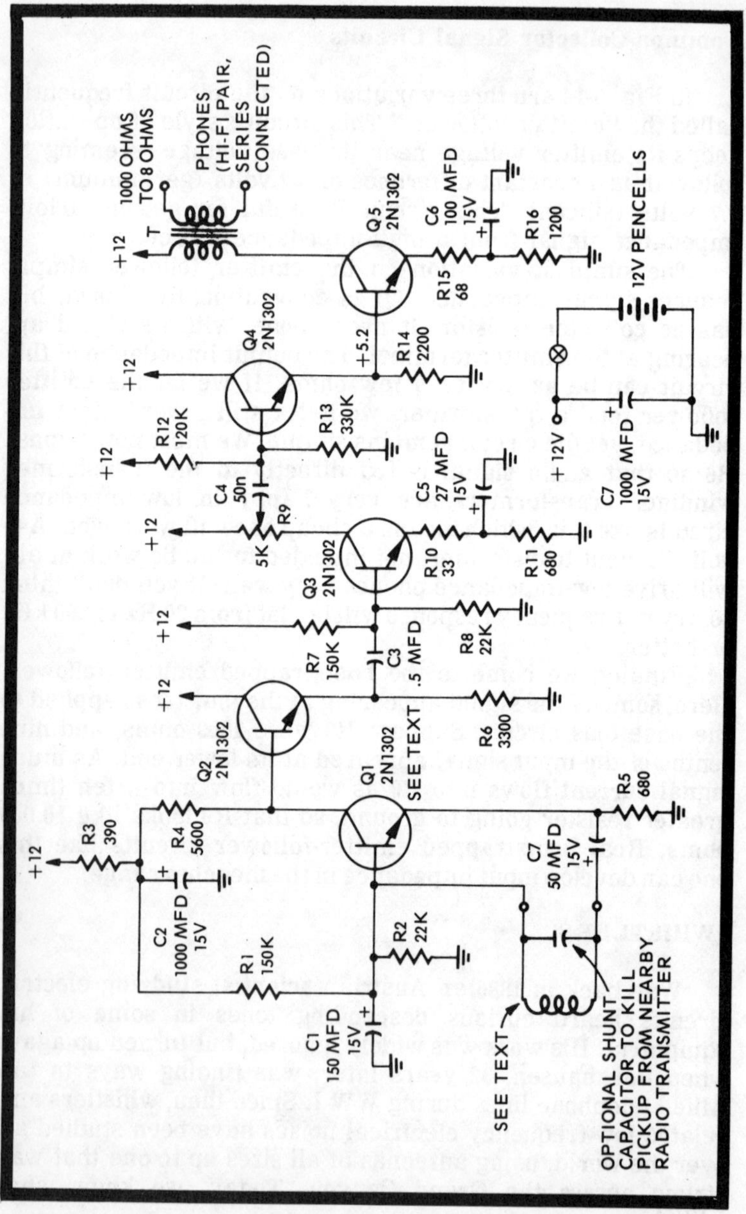

Fig. 6-15. This very high-gain audio amplifier should not be built in too small a space, but no shielding was needed in the original version. Input stage is decoupled to improve overall stability.

force between the Earth's poles. In the process, high-frequency components of the static travel faster than the low-frequency components and arrive first. The lower frequencies tag along after, giving rise to a descending-tone hiss resembling a whistle. Since whistler signals follow a definite route, they are only audible to a listener in the Northern hemisphere when a thunderstorm is going on at his matching region in the Southern hemisphere. This probably is why it took about 80 years to find what whistlers are.

Other electrical events cause the "dawn chorus," a remarkable set of noises sometimes heard shortly before sunrise, and "hiss," which originates in the ionosphere. All that is needed to hear these sounds is a high-gain amplifier fed by a large pickup loop. Since they are already at audio frequencies (20 Hz to 20 kHz) no radio receiver is needed; the electronics only has to amplify the signal enough for earphone listening. This makes for some very simple electronics, but the signals are very weak. They must win over AC interference, natural static, local interference such as radio stations and radar sets, and circuit noise. You cannot listen for whistlers within hundreds or thousands of feet of house wiring

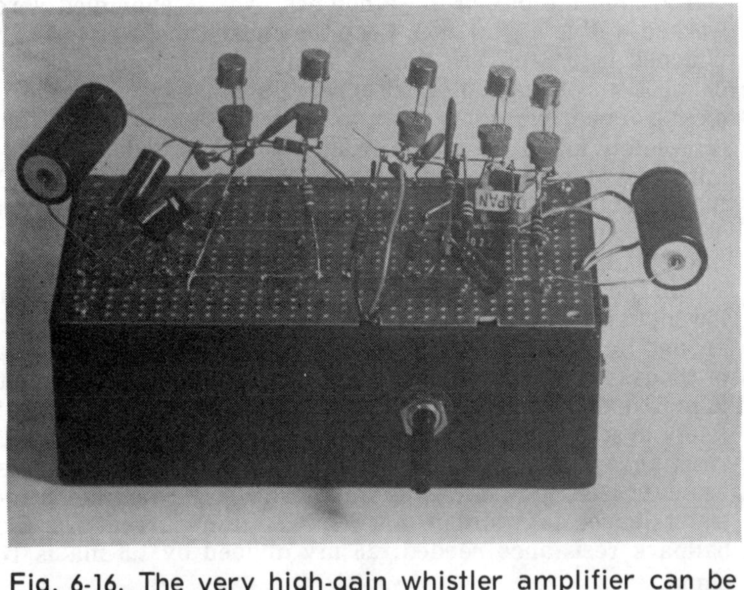

Fig. 6-16. The very high-gain whistler amplifier can be constructed breadboard style in about a half hour. Notice carefully used large decoupling capacitors.

and AC power lines. And once all is right you still have to wait for that thunderstorm in the right area. Just hearing a whistler is a challenge, and it takes some engineering to meet it.

Since all charge flow is a bit grainy in structure, a little like pouring gravel, electric current makes noise. Whistlers and related signals must win over this noise to be heard. Two approaches work. One is to use a very large antenna, so the incoming signal is stronger. And the other is to use very quiet circuits for less interference. The engineer thinking along these lines is said to be seeking a good "signal-to-noise ratio." The amplifier described here is designed along these lines, but with a requirement that it cannot use special hard-to-find parts. The circuit in Fig. 6-15 manages to bring an estimated 24-nanovolt (.024 microvolts) 800-Hz signal to audibility within strong noise. A 240-nanovolt signal comes through well. Probably this performance can be improved, and we will touch on that later. The completed unit is shown in Fig. 6-16.

The amplifier has to have a loop to pick up the incoming signal. To make a usable loop, begin with a large but convenient sized cardboard box with a good bottom. Cut off the box an inch and a half to two inches above the bottom, leaving a flat pan affair. Notch each corner and wind the wire in a big coil around the box in these notches. No. 26 enameled wire worked well in a trial loop. Loop the ends of the wire through punched holes and bring in to a miniature phone jack for connections. A capacitor wired across the jack will reduce pickup of nearby radio transmitter signals. A 1-mfd capacitor completely killed the signal from a 50-KW broadcast transmitter less than a mile away. Since the loop is sensitive to all magnetic fields, including that generated by the amplifier, it should be placed several feet away from the amplifier. Lapel microphone cable is convenient for interconnection.

In the circuit, Q1 is a common-base amplifier biased for low input resistance. Few of us could make a pickup loop large enough for a Thevenin equivalent resistance in the hundreds or thousands of ohms; therefore, a common-base input circuit is most effective for the kind of input this amplifier is most likely to see. A convenient-to-make loop proved to have a DC resistance of 14 ohms, and at audio frequencies its inductive reactance is too small to be important. A germanium transistor biased to maintain a 1.5 ma emitter current has the ballpark resistance needed: 26 mv divided by 1.5 ma is 17 ohms.

We choose an input resistance near the loop's Thevenin equivalent because this transfers the most power into the amplifier. This condition is sometimes confused with maximum efficiency, which requires a load resistance as small as possible compared to the source's Thevenin equivalent resistance. Notice we have a common-emitter bias circuit combined with our common-base signal circuit, with the base signal-grounded through C1. This arrangement is convenient for RF applications, too.

The Q2 emitter follower minimizes the loading on the Q1 collector for maximum voltage gain in Q1. Alternating gain stages and emitter followers is a nice way to simplify elementary circuit design. The Q2 output goes through C3 to common-emitter amplifier Q3. Unbypassed resistor R9 in the Q3 emitter circuit reduces circuit dependence on transistor characteristics. A gain control, R8, serves as the Q3 collector resistor, with the audio signal going to another emitter follower. Finally, audio output transistor Q5, biased near 5 ma current, provides audio to the phones through transformer T1. The phones are a pair of hi-fi headphones connected in series by appropriate wiring to a 3-terminal standard hi-fi jack. Hi-fi phones introduce practically no signal distortion or intermodulation; they mask outside noise, and offer very good sensitivity. They were used with this amplifier and are recommended over any loudspeaker for all lab work.

Upon completion, this amplifier proved unstable. The professionally employed engineer faced with instability must solve it the least expensive way, but we have more freedom. A 1000-mfd capacitor across the battery, another in a decoupling network for Q1 and Q2, and disc ceramic capacitors to ground from the Q1 and Q3 collector terminals were added before the next test. The amplifier now proved stable with an open-circuit input.

Turning up the gain control produces strong noise, which originates in the input circuits and transistor. It is the natural noise mentioned earlier, present in all parts of the circuit. Since only the input circuit sees the entire input-output gain, its noise is tens or hundreds of times more noticeable than any other circuit-generated noise. Replacing the composition resistors used in the Q1 base and emitter circuits with metal-film or wirewound resistors should reduce noise noticeably, but attention in the breadboard went to choosing a better transistor. This proved worthwhile.

Two transistor types which reduce output noise are quiet and not-working. An audio VTVM at the amplifier's output gave readings with no signal input for various transistors. Each reading was written down and the associated transistor placed beside it; results accumulated for six transistors showed the two best ones. One went permanently into the Q1 site; the next best into Q2. Then a sensitivity test proved neither transistor was low-noise simply because it didn't work.

Circuit performance is verified by feeding it an input resembling its intended input. A simple network of an 82K and a 10-ohm resistor reduces a typical tens-of-millivolts audio oscillator output to the needed submicrovolt levels. This was used to get the voltage sensitivity result mentioned in the opening paragraphs. Unfortunately, this circuit's ultimate performance could not be proven by real-life tests. AC interference, New England winter weather, and time pressures prevented completion of a true performance test. But it's the right start and has passed realistic, tough bench tests. Can you finish it?

A DIODE RECEIVER

Radio history has it beginners and diode receivers are closely associated. Building a diode receiver is supposed to teach you something, although it is never very clear just what. The basic circuit is too simple to have much teaching value. It isn't up to sorting out multiple nearby transmitters, and needs a large antenna. And the circuit usually appears as a collection of parts assembled without regard to any engineering principles. More details come to mind: If the diode receiver has a bad reputation, its designers earned that. Yet, if we do sit down and try some engineering it's possible to come up with a circuit that is simple, educational, and which also has real utility value.

Since the basic principles are covered all through the book we'll hit only the high points here. Probably the most important of these concern the diode detector. See Fig. 6-17. Ideally, we want the diode not to detect any signal at all, since the detected audio signal is actually energy stolen from the tuned circuit (L1-C1). This is the only tuned circuit in the receiver and the more energy lost from it, the poorer the selectivity. In the early days of radio this made little difference, but in today's conditions there are often several transmitters nearby in a crowded metropolitan area.

If the detector circuit is to absorb little energy, it must feed a high resistance. Neither "high-resistance" phones or a simple transistor circuit have much going for them in this department. Both loads are in the low 1000s of ohms—too low. And the poor efficiency of either makes it an outstandingly weak candidate for DX (long-distance) listening. This is alright if you live on acres of land since you can set up a large antenna and dig a good ground. But today's beginner needs a

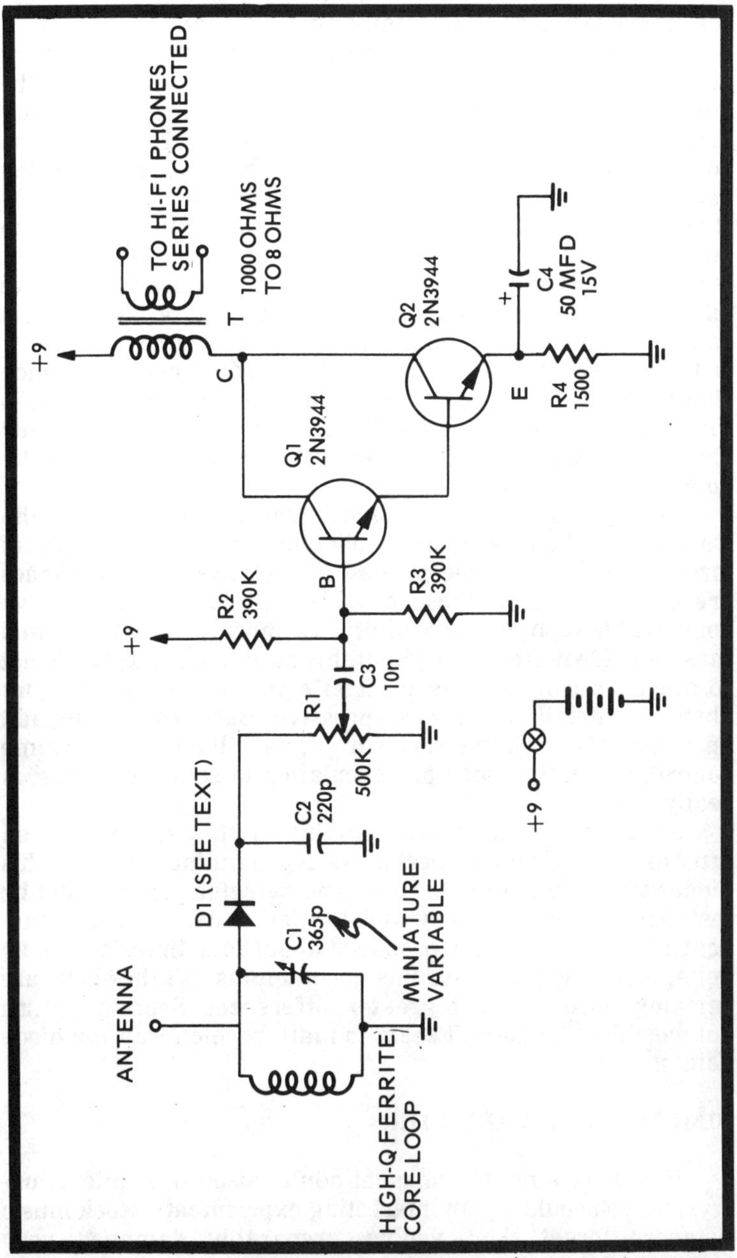

Fig. 6-17. A little engineering improves the basic diode receiver to a point where its performance is both instructive and useful.

circuit with more performance potential—and a high-resistance load for the detector is one key point.

Even making the best of all that, we still have fairly grim design requirements. Except under rare conditions, the detected signal is in the low-millivolt range, according to my breadboard tests. What **simple** circuit of easily-found components will amplify these very weak signals and have the necessary high input resistance? A Darlington amplifier will. It uses two transistors wired as a single super-transistor. The points labeled E, B, and C act as emitter, base, and collector leads of the Darlington pair, Q1 and Q2. Because Q1 acts like an emitter follower, this super-transistor has a higher input resistance and power gain than any commonly available ordinary transistors. Its roughly 120K input resistance (ballpark) is 100 times greater than either transistor alone in the same common-emitter circuit. With more complicated electronics, we could do better but (circuit designers take note!) this works well enough.

Finally, an inexpensive audio transformer matches the Darlington's high-resistance collector circuit into a pair of low-resistance hi-fi type phones. Short-wave listener phones are out the window these days because hi-fi phones have comparable or better sensitivity, fit more comfortably, and mask outside noise. Throughout this circuit our purpose is not so much "sensitivity" as **most effective design.** Each point above is a detail—not very impressive. But these details add up to a really workable receiver. Tests will show the volume control, evidently added on speculation in some receivers, is really needed.

An important unknown remains in this receiver, even after assembly from the best parts. No manufacturer rates his diodes for circuits like this one. The detector diode should be installed in clips for easy substitution tests. During breadboard tests some diodes produced about four times the audio voltage of others, and surplus germaniums, off-the-shelf, are certainly good for much greater differences. Some may turn out worthless. It doesn't seem to matter which way the diode points.

SIMPLE HI-FI AMPLIFIER

If you are a not too unusual adult raised in a quiet countryside, you could try an interesting experiment. (Rock music listeners forget this; you've irreparably damaged your hearing.) Some young people's hearing sensitivity approaches a theoretical limit, and if yours is really good, maybe you can reach it. If so you can hear the thermal hiss of air molecules at a sound power level near the generally accepted threshold of hearing. But you must have true silence for that, so probably you cannot do the experiment. If you can, and you succeed, that noise is the same kind of noise found at the whistler amplifier's output. It is a result of our universe's grainy structure and continuous thermal agitation.

Average human hearing sensitivity, near this limit, is the "zero decibel" power level referred to when a sound level meter reads 60 db (a quiet room, one million times noisier than threshold of hearing) or 120 db (a million times louder than the quiet room; hearing damage in a few minutes.) Its power level is 10^{-12} watts per square meter. This surprising number brings high power hi-fi instantly to a new focus. Do we **really** need power levels of a thousand million to two hundred million million times our hearing sensitivity to enjoy high fidelity listening? That seems unlikely.

Other facts suggest high power levels are far less needed than the hi-fi advertising literature indicates. A man talking generates about 50 microwatts of audio. A large pipe organ is good for about 3 milliwatts. Against these sound power levels, 30-, 100- and 200-watt hi-fi amplifiers seem far out of scale—and they are. We just don't need powerful electronics for hi-fi, apart from practical considerations we may not care to spend the money, upset our neighbors, and damage our hearing. A pair of hi-fi headphones will mask outside noises (far too common today, and often ugly) and do away with several tough engineering problems.

In the circuit in Fig. 6-18, common-emitter amplifier Q1, with a gain control in its collector circuit, feeds common-emitter amplifier Q2. Emitter follower Q3 feeds emitter follower Q4, which supplies audio to one side of a hi-fi headphone pair through transformer T1. We use two emitter followers in series so the load resistance referred back to the Q2 collector is seen multiplied by the betas of two transistors. It is so large it can be ignored compared to R11, which eases the choice of values for C5 and R12. We come back to C5 and R12 shortly.

Output transformer T1 is a common 6.3-volt filament transformer. It works fine in low-impedance circuits, which we have here. And the transformer relieves Q4 of delivering really large currents, which it cannot do well. A cheap filament transformer is no quality audio device by any stretch of the imagination, but in this circuit you cannot find the difference.

It's easy to suppose all circuits return to ground in any electronic system. Where is ground in this one? Its input is

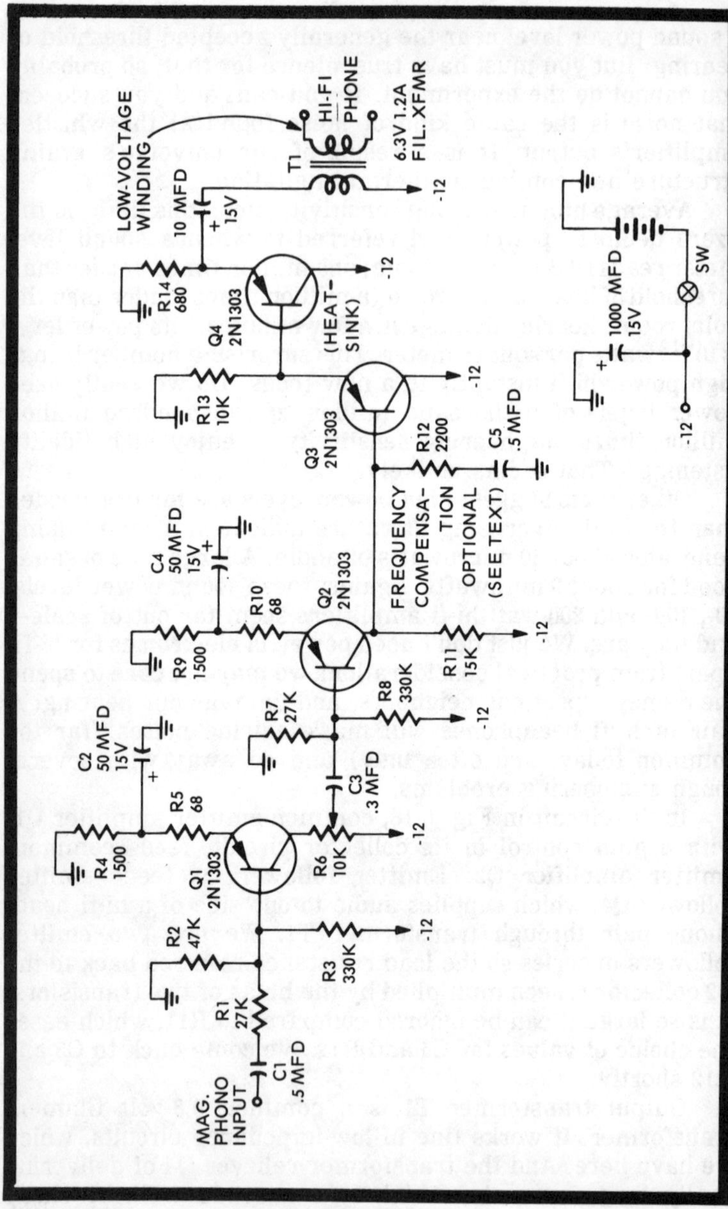

Fig. 6-18. A few inexpensive components can be assembled into an uncritical phono amplifier whose performance is surprisingly good. Two of these are needed for stereo, of course.

referred to the positive supply line as ground. But its output end uses emitter followers that see the negative supply line as ground. That is why the T1 primary returns to the minus line. Missing this detail during circuit development wasted time chasing a persistent instability.

Q4 is heavily biased so it can turn on for lots of muscle during negative-going signal swings, but for positive-going swings performance is limited by what a 680-ohm resistor can manage. This is hardly adequate, which is why Q4 is biased to carry lots of current. Q4 might be heatsinked, if you like to be careful. Also, you might try a Class A complementary emitter-follower here, as an alternative to the design shown.

RC network C5 and R12 serves as frequency compensation. The record manufacturers de-emphasize low frequencies in recording (RIAA characteristic) to reduce track width and attendant problems on their discs. C5 and R12 act as load on the Q2 collector circuit to reduce the HF response, which is another way of building up the lows. At low frequencies C5 hides R12 from the Q2 collector terminal; at high frequencies R12 appears as an additional load, considerably reducing the Q2 stage gain. Since Q3 and Q4 isolate the heavy output load from the Q2 collector terminal and Q2's output resistance is in the hundreds of thousands of ohms, the Thevenin equivalent of this circuit is a voltage source with an output resistance equal to R11 only. This makes the choice of C5 and R12 easier. Choosing a larger R12 strengthens the highs and brings the HF roll-off knee to a lower frequency. And choosing a larger C5 lowers the HF roll-off knee but does not change HF attenuation. These facts make tone control a little tough; an elaboration of this circuit might be a switch and three or four choices for C5-R12 combinations. The values given here are a workable beginning.

Small resistors in the Q1 and Q2 emitter circuits improve linearity. There is no definite indication they are needed—or aren't. Shorting them increases volume slightly. Probably it increases signal distortion, too. Finally, series resistor R1 in the input circuit increases the resistive load seen by the magnetic phono cartridge. This comes out 40K estimated; near enough to the generally accepted 50K figure.

This amplifier is simple and small enough to fit inside a hi-fi turntable (Fig. 6-19), but it will take some woodworking. New turntables are expensive; some audio shops deal in used gear which will prove entirely satisfactory if sensibly chosen. Here are some hints.

A changer that cannot change makes a perfectly satisfactory turntable. A good magnetic cartridge should be

Fig. 6-19. This high-fidelity amplifier uses a common 6-volt filament transformer for matching its output to a hi-fi headphone. Since the circuit is designed for it, the filament transformer works fine.

used, one that tracks at three grams or less. Higher tracking pressures wear disc uneconomically fast. A bad diamond or sapphire needle ruins good records in one playing; the needle should be checked under a microscope and replaced when at all worn. If you use this amplifier you will want to keep your records extremely clean. These simple electronics put out a very pure signal (but maybe the HF compensation needs further adjusting). If hum, static, or clicking noises appear check the connection from the turntable metal base to the plus side of the electronics. At 10 ma drain, D size cells will prove more economical than pencells.

Finally, the choice of hi-fi headphones is important. Advertising literature or a store salesman are no help, at best. Since the market changes fast, a wise step is reference to a publication like Consumer Reports. At this writing Allied Radio and Radio Shack market a hi-fi headset called their "Nova Pro." This features truly great bass response, good treble response, and very acceptable styling. Its headband could use work since the inside steel band is too narrow. But one year earlier they were selling an entirely different and less satisfactory headset under the same name. Look out for facts in the hi-fi market—if you can find any.

Chapter 7

Integrated Circuits

Electronic circuits ranging from transistor pairs to complete computers are available now as tiny silicon "dice" up to a quarter-inch square by a few thousandths inch thick. These are called "integrated circuits," or "ICs." From a circuit viewpoint, ICs contain a few real innovations, but they are all-around better (excepting a minor lack of flexibility) than circuits assembled from discrete components. But the many detail differences between IC electronics and discrete components, taken all together, come near a difference in kind. The result is practically a new electronics technology.

An IC circuit is simply a piece of silicon crystal, subtly but remarkably changed by certain industrial processes evolved from transistor construction technology. Where a transistor is little "spots" of selectively doped crystal designed for transistor action alone, the IC also contains resistors and capacitors, diodes and zeners, interconnections, and special circuits that act like large inductances and capacitances. Some ICs go beyond even that and include solid-state mechanical structures for tuning and control purposes.

The equipment needed to make ICs is very expensive and extremely precise. Usually, it is automated. This boosts capital investment, reduces operating costs, and improves uniformity over the many fabrication processes. All this capital and operating cost must be carried by the product, but a single production run on a few tens of 2-inch diameter silicon crystal slices makes hundreds per slice of practically identical ICs. The production cost per dice turns out small enough that complex ICs are available at dollar-level prices from manufacturers and retail dealers, and at much lower prices in the surplus market.

What is conventionally sold as an IC is actually a package made large enough to handle easily. A few percent of this package's weight is actually the IC; the rest is a block of plastic, a metal header and cover, or a sealed ceramic assembly that protects the working dice from the atmosphere. This block or container also supplies the needed stiffness for mounting rugged outside leads while maintaining very

Fig. 7-1. A striking illustration of miniaturization—two T05 transistors (2N1302), a PA230 integrated circuit amplifier (lower left) and a vacuum tube of a type popular a quarter century ago.

delicate connections into the real IC dice. If the IC is a power switch or an amplifier, the block also contains some provision for conducting waste heat from the IC chip to a nearby sink. Standard practice in modern electronics is to speak of the silicon dice IC and the overall assembled working package by the same term: integrated circuit, or IC.

Very important technical advantages evolve from IC technology. Here are a few; others interest workers more concerned than we are with engineering. ICs are tens to thousands of times more reliable than similar circuits built of discrete components, as well as being tens to hundreds of times smaller. They are ten-thousands to millions of times smaller than the vacuum-tube circuits that were made obsolete shortly after WW II, if you like to think about history. ICs use tens to hundreds of times less power than discrete component circuits, and weigh less (important for space exploration). Because their smaller electrical structures work faster and signals travel shorter distances inside, ICs work at higher frequencies or do computing and control work far quicker than discrete component circuits. Any one of those advantages would give ICs a strong edge over anything in discrete component technology. Taken all together, they explain why modern electronics is very much an IC systems technology.

The following treatment of ICs isn't a "patch" to what the subject deserves. If that's true, why write about discrete components at all? Because discrete component circuits emphasize basic principles we cannot get away from and because, alas, you cannot reach inside an IC and make changes. Sometimes you must take pencil and paper, begin from scratch and the parts bin, and originate that circuit for what nothing in the catalog can manage.

From all this we can make a very good guess what future years must bring in electronics technology. We will more and more use ICs, usually with some discrete components tacked on the outside. This won't relieve us of any challenging circuit design problems. The manufacturers know that perfectly and often supply extra terminals connected deep into their ICs for special adaptation connections. This improves utility—and market-ability. Amateur electronics can become simpler for those who like it that way, but many of us will find the new IC-oriented technology livelier, more versatile, and more challenging than any of its preceding technologies.

THE GENERAL ELECTRIC PA230 LOW-LEVEL AMPLIFIER

The PA230 is a little black, flat block three quarters of an inch long. See Fig. 7-1. A tiny chip inside carries eight transistors, three diodes and a zener, and 15 resistors, along with interconnections to make up a working amplifier circuit. Since the manufacturer has managed parts sizes, matching, and interconnections, little remains to do in applying the PA230. When we have some idea what we want to try, we can come up quickly with a working if less than beautiful breadboard. And a finished "frozen style" project using the PA230 can be done in a fraction of the time needed to assemble a comparable device from discrete components.

Fig. 7-2 shows us what's in the PA230. We see its input is a cascaded difference amplifier feeding an emitter follower, Q6. This, in turn, drives output circuit Q7 and Q8, which is a very puzzling circuit. The following discussion didn't originate at G.E. Evidently, the two diodes in the Q8 collector circuit have quite different properties, allowing Q7 to work as an emitter follower under normal conditions. This would account for the PA230 output resistance spec of 100-200 ohms, which is about right for a very low-current emitter follower (silicon tran-

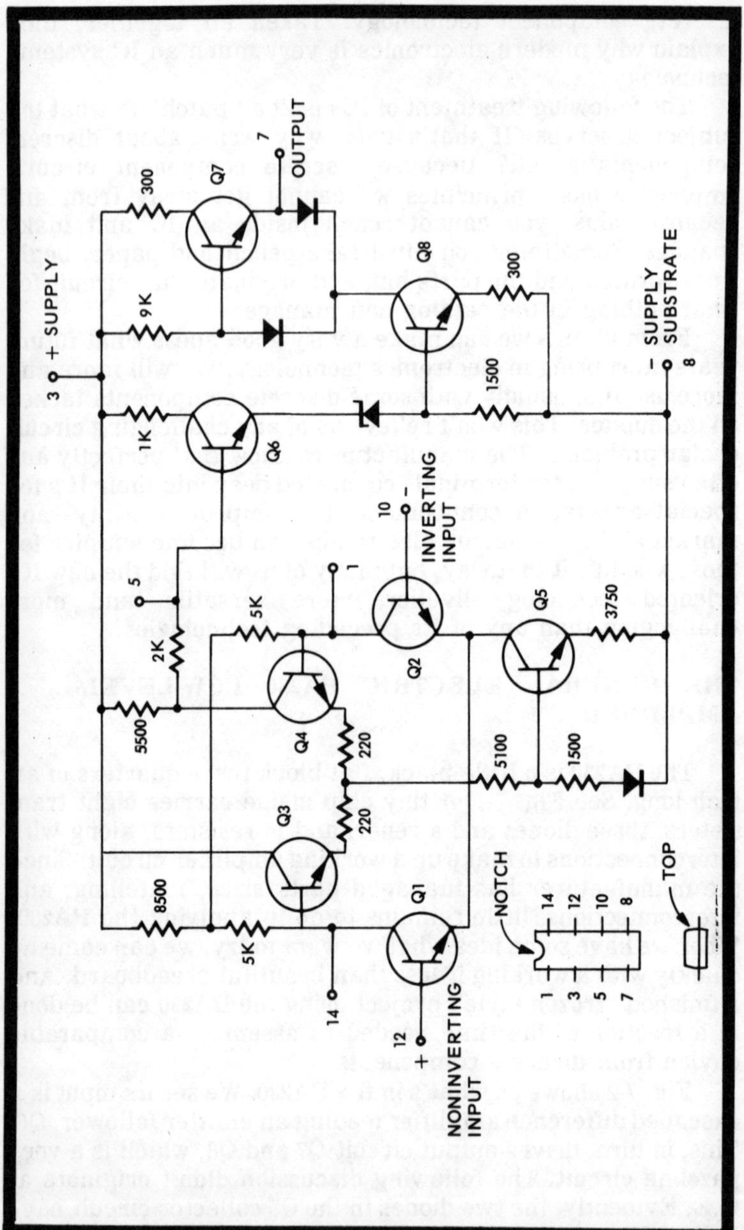

Fig. 7-2. The PA230s tiny silicon chip carries this complex amplifier circuit.

sistor running 0.2-0.3 ma emitter current). But a very large signal reduces voltage from the Q7 base to the Q8 collector, tending to turn off Q7, which avoids its destruction from overheating. Maybe that's it, and maybe not; at any rate it's an odd circuit.

Earlier, we were concerned with the simplest basic circuits. Now we face a new one—the difference amplifier. See Fig. 7-3 consisting of 2N1302 transistors Q1 and Q2 and a few resistors and capacitors. Q1 and Q2 are biased through the large emitter resistor, R2. Capacitors C1, C2 and C3 simply block DC voltages; they don't take part in circuit operation. The Q1 and Q2 base currents enter through R1 and R4, and Q2 collector resistor R3 serves to convert the Q2 collector current signal into an output voltage signal (Ohm's Law).

We assemble this circuit and check bias voltages. If Q1 and Q2 are sharing the R2 current equally, the Q2 collector voltage will be about 3 volts. (6 ma current splits to 3 ma per transistor; 3 ma produces a 3-volt drop across R3.) If this voltage is low or high, Q1 and Q2 betas are too different. Leaving Q1 in its socket, we choose Q2's for a ballpark 3 volts

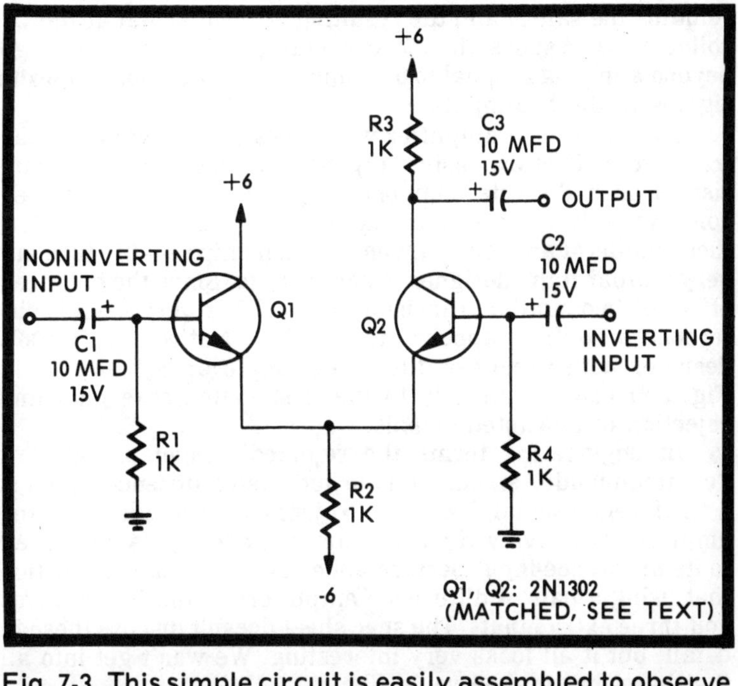

Fig. 7-3. This simple circuit is easily assembled to observe the basic properties of difference amplifiers. Probably the transistors must be matched for similar betas.

collector voltage. In linear operation a transistor's collector current is little affected by its collector voltage, so the resistor is omitted from the Q1 collector circuit without appreciably influencing transistor current sharing.

Now let us see what happens if we apply the same signal to inputs "+" and "-." We put a clip lead across the two input terminals and apply perhaps 10 millivolts audio to this circuit. The result is practically no audio output. Why? It is because neither overall transistor current or the current sharing are appreciably influenced by the incoming signal. Sharing isn't affected because both bases see the same signal. And total current isn't affected because the 10 millivolts of audio, applied across R2, influences voltage and, therefore, current by 10 millivolts in 5.8 volts, or about one part in 5800. The difference amplifier is very insensitive to signals applied to **both** terminals simultaneously.

Now we ground one input and apply our 10 millivolts to the other. We discover maybe 140 millivolts at the output terminal now—a voltage gain of 14. Now the current sharing between Q1 and Q2 is being changed by the signal (average current remains the same) and the resulting current signal at the Q2 collector is responsible for our output. We could even go beyond applying a signal to one input; we could apply **opposite** signals to the two inputs.

The difference amplifier works best, using very similar transistors. It is expensive to pick and choose similar transistors on a discrete components production line, but they come very easily in IC making. Two transistors made a few thousandths of an inch apart on a silicon chip reliably turn out very similar if the designer wants them to. Since the balanced effect of two similar transistors in an IC is very useful, difference amplifiers appear often in ICs. Notice the PA230's designer added another difference amplifier (Q3 and Q4 in Fig. 7-2) coupled directly to the first to increase gain and rejection of unwanted signals.

In engineering terms the rejected 2-input signal is a "common-mode" signal, and the accepted difference signal is a "difference signal." A large "rejection ratio" indicates the amplifier effectively rejects common-mode inputs compared to its difference-input performance. Circuit designers, notice that using a pair of difference amplifiers in the PA230 gives you three extra inputs. The spec sheet doesn't discuss these in detail, but it all looks very interesting. We won't get into all that here, but maybe you'd like to.

An additional terminal, Pin 5, appears as either input or output. We will use this later for a little paring of the PA230 high-frequency response. Like most ICs, the PA230 has very small transistors, giving it much better high-frequency response than is needed for most purposes. Sometimes this HF response causes trouble.

Finally, see Fig. 7-4 which is a table of PA230 specifications from G.E. For critical applications, check recent G.E. product literature. Notice this table doesn't mention the properties of Terminals 14, 1, and 5. The remedy is simple: we study the schematic, work out an idea we'd like to try, and then we quick-and-dirty breadboard our idea to see what happens. How soon can you breadboard a circuit with eight transistors, three diodes and a zener, and 15 resistors? Using the PA230, 10 minutes should do it. Just enough time to warm up your test gear.

A DECADE AMPLIFIER

Instrument sensitivity is a perennial amateur electronics problem. The gear most amateurs can afford doesn't feature extreme sensitivity. Typically, an input of one-tenth to one volt to a scope or meter gives a usable reading. Smaller signals such as microphone outputs and low-level communications signals are hard to measure or observe. The PA230 provides a remedy for this problem, as shown in Fig. 7-5. The sidewise triangle is the conventional symbol for a high-gain amplifier.

	MIN.	TYPICAL	MAX.
OPEN-LOOP VOLTAGE GAIN	4,000	7,000	12,000
SUPPLY CURRENT		3 MA	5 MA
INPUT BIAS CURRENT		2 MICROAMP A	
OPEN-LOOP BANDWIDTH		500 kHz	
INPUT RESISTANCE	15K	35K	
OUTPUT RESISTANCE		100 OHMS	200 OHMS
OUTPUT VOLTAGE SWING		10V P-P	
SUPPLY VOLTAGE	10V	12V	15V

Fig. 7-4. PA230 specifications (from General Electric).

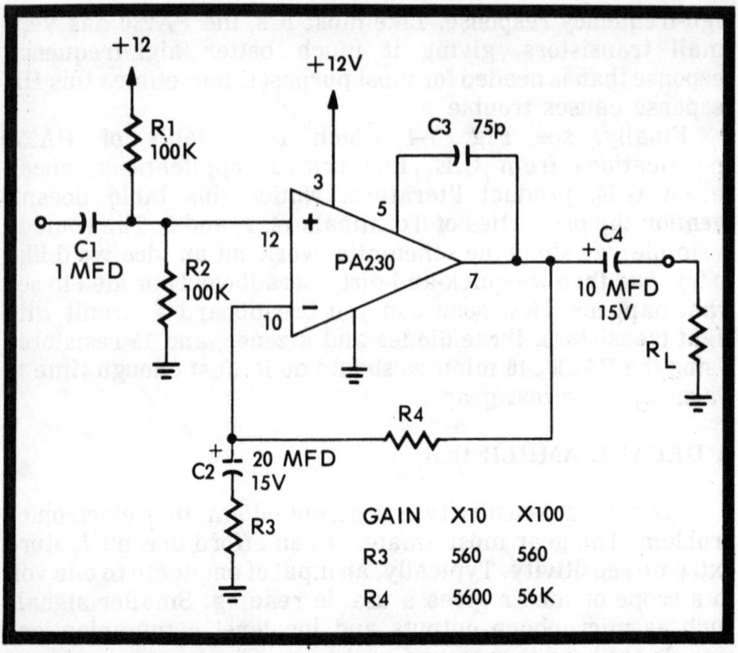

Fig. 7-5. Schematic of the PA230 as a decade (input times ten-factors) amplifier. Where the input and output currents flow is extremely important in applying this high-gain amplifier.

The PA230 gain of 7 to 12 thousand doesn't appear between input and output terminals because R4 and R3, feeding back some of the output, adjust the apparent gain very near G equals R4 divided by R3. This is discussed later.

Fig. 7-5 is a deceptively simple schematic, but let's trace the signal path. Input goes to Terminal 12, with a DC bias supplied by R1 and R2. Since the PA230 gain is very large, a strong signal appears at Terminal 7 to be distributed to three circuits. We begin to see why this schematic is not so simple. Some signal returns to Terminal 5 through an 82-picofarad capacitor. Low-frequency feedback through C3 is nearly zero, but an inspection of Fig. 7-2 suggests the return signal comes into about a 3000-ohm load. The capacitor reactance is 3000 ohms at about 700 kHz, placing circuit response roll-off at a somewhat lower frequency.

The second signal circuit proceeds to the output load; no adventure there. Now we come to it. What about the signal entering the circuit of R4 and beyond? This signal returns to

Input 10, labelled "-," through the voltage divider (R4 divided by R3). The "-" means any signal to this input opposes that to the "+" input (12). The schematic in Fig. 7-2 shows us these inputs offer equal gain through the amplifier. The minus return input then acts to cancel any plus terminal input from the outside world.

The amplifier lives with this apparent contradiction and it is greatly improved by it. It turns out the overall effective voltage gain freezes at the ratio R4 over R3. Actually there is some error, but it is small if the controlled gain is not too much of the amplifier gain. In other words, if the PA230 gain is 7000 we should not try to approach that gain too closely. This is why the schematic carries a warning that the circuit's gain will have to be zeroed in on 1000. Since resistors R3 and R4 fix effective amplifier gain, we can achieve a certain performance without doing special tests to prove or adjust it. We just choose very good resistors for R3 and R4. The best choice is metal-film 1 percent types, but 5 percent or even 10 percent resistors will serve for breadboarding. Wirewound resistors aren't appropriate here because of their slight parasitic inductance.

Finally, capacitor C2 prevents the DC output level at Pin 7 from being lost to ground. The difference amplifier input needs appropriate DC as well as signal voltage inputs. And C2 reactance rises at low frequencies, which reduces the amplifier gain. The 20-mfd value suggested puts the low-frequency roll-off near 20 Hz. Remember, electrolytic capacitor tolerances are typically loose, possibly -20 to +100 percent.

To avoid perplexing trouble with this circuit, we have to think about where input, output, and miscellaneous loop currents will flow. Our input loop is through Pin 12, out Pin 8, and back to the input source. Our output loop is Pin 7, to load, and return through Pin 8. Our minus supply loop also goes to Pin 8, and the feedback network and input biasing connect here. Evidently we had better respect Pin 8.

One ground and bypass arrangement has a small wire V at Pin 8. The side toward Pins 10 and 12 is devoted to input and feedback network grounding. The other end is the Pin 7 return and power supply connections. A 10-nanofarad or larger disc ceramic is connected between Pins 8 and 3 for high-frequency bypassing. A larger electrolytic capacitor nearby may or may not turn out to improve circuit operation.

Suggestions about stability are very much in order at this point. Whenever working with a linear IC, assume it is oscillating and keep looking for the reason until you've proved it's stable. A voltage gain of thousands or ten-thousands across a silicon chip less than fifty thousandths across has troublemaking potential you mustn't underestimate. The PA230 circuit described here tended to oscillate at 10 and 1000 gains until its output (floating into a high-resistance scope probe) was loaded to 10K and its feedback circuit (R4 and R3) was boosted to higher resistance levels than those first chosen.

To stabilize jumpy ICs, first check connections to ensure real connections match those on the schematic. A possibly crummy capacitor should be replaced. Appropriate bypassing of supply leads becomes attractive if trouble appears (might as well do it right the first time around!). A feedback network can be tinkered with, raising or lowering its resistance levels without changing feedback ratios. A tiny capacitor (called a phase lead capacitor) of a few picofarads across the return resistor (R4 in Fig. 7-5) will do something remarkable. Maybe it will be good. Input bias resistors can be changed to boost or decrease the resistive load they present to the input circuit. And an unruly IC acquired from the surplus market may have been there because it was unruly. Maximum PA230 current drain is 5 ma at 12 volts. An ordinary pencell at this level for 4 hours per day is good for about a thousand milliampere-hours. Estimated battery life is 60 days.

THE uA709 OP-AMP

A useful new member of the basic electronics components set is the operational amplifier. Inside, it is a complex, high-grade direct-coupled amplifier, but we are often no more concerned with such details than we are interested in the materials that make up resistors, capacitors, inductors, diodes, and so on. The op-amp emphasized here is the Fairchild 709C, but others are about the same, once basic power and frequency compensation connections are completed.

Operational amplifiers are very important to anybody concerned with modern electronics. They are simple enough on the outside and work in many apparently simple circuits. They amplify signals, control ovens, emphasizes or filter frequencies and signals, drive headphones, generate specially shaped waveforms, work in computers—all with equal effectiveness. But it's necessary to understand them well enough to avoid trouble. The 709, a typical op-amp, contains 14 transistors and 15 resistors, and its gain extends to high frequencies. It is a natural for converting bad engineering into bad results. For an overall picture of the 709's characteristics, see Fig. 7-6.

	MIN.	TYPICAL	MAX.
INPUT OFFSET VOLTAGE		1.0V	5.0V
INPUT BIAS CURRENT		.2 MICROAMP	.5 MICROAMP
INPUT RESISTANCE	150K	400K	
OUTPUT RESISTANCE		150 OHMS	
LARGE-SIGNAL GAIN	25K	40K	70K
SUPPLY VOLTAGE			+18V (EQUAL)
DIFFERENTIAL INPUT			± 5V

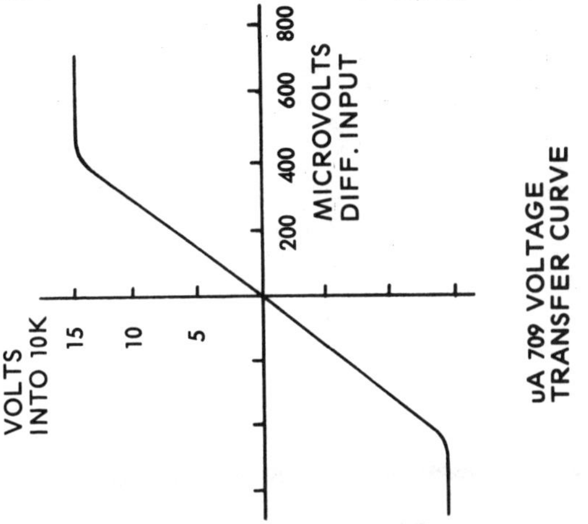

Fig. 7-6. Overall characteristics and transfer curve for the Fairchild uA709 operational amplifier.

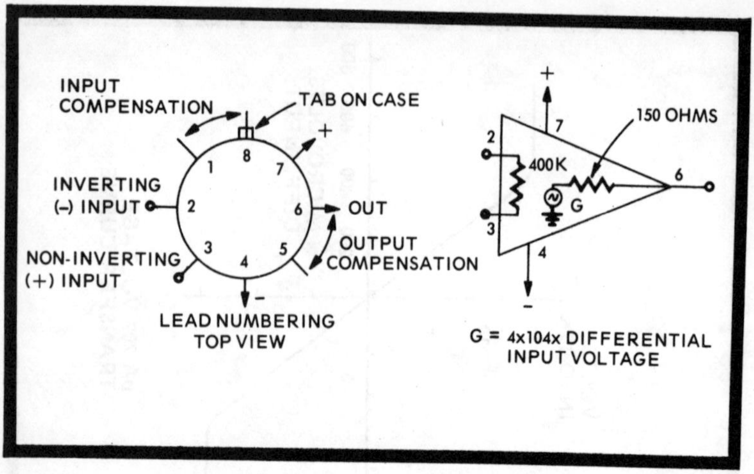

Fig. 7-7. uA709 terminal connections. The equivalent circuit summarizes the most important characteristics of the 709.

The transfer curve shows what "high gain" really means. For differential inputs (between + and - terminals) exceeding about 300 microvolts, the output clips. This detail has immense practical importance, which appears in a rather curious way. And the bandwidth concerns us, too. The major part of avoiding problems is to reduce and control this bandwidth. Fairchild's designers were aware of this need when they originally designed the 709 (the first op-amps used vacuum tubes but had about the same good and bad points as later solid-state op-amps) and added the appropriate terminals for compensation (stabilization) networks.

In Fig. 7-7 we find the 709C's characteristics represented by an equivalent circuit. Supply and frequency compensation connections are usually assumed, as here. The two inputs work independently of ground, with an output voltage proportional to amplifier gain times signal across the inputs. Notice the resistance between the input terminals is about 400K, a large but not infinite value, and the output comes from a Thevenin equivalent whose resistance is about 100 ohms.

Rules Of Thumb

We can avoid mathematical analysis of the op-amp's operation by making some connom-sense simplifications of the usual specs:

1. We act as if the op-amp input circuit draws zero current. Actually, it draws a little, but this is so small compared to signal currents nearby that we can usually ignore it.

2. We act as if the op-amp responds to a signal too small to look different from zero. This is less far out than it seems. In the 709, an input to generate a given output is something like a half thousandth of 1 percent of it. We aren't far off calling that zero.

3. Sometimes it helps the math to suppose the op-amp's voltage gain is infinite. A mathematician may point out 15,000 is a pretty weak guess at infinity, but electronic circuits see things differently than we do.

Later we will apply these rules to develop explanations of how two basic op-amp circuits work.

Practical Application Details

At this point it seems logical to look at applications. We had better not, since the op-amp has some tricks we have to tame. We first look at a peculiar kind of runaway called "latching." See Fig. 7-8 illustrating a 709C latched hard-on.

This happened because some accidental event or design weakness allowed or forced the 709's minus input transistor into saturation. That allowed the input to bypass this transistor's phase inversion and reach the next transistor further into the 709's circuits; the minus input now appears as a plus or regenerative input. The already positive input is driven further positive if it has not already reached its limit and if there is a DC connection (no series capacitor) from output to

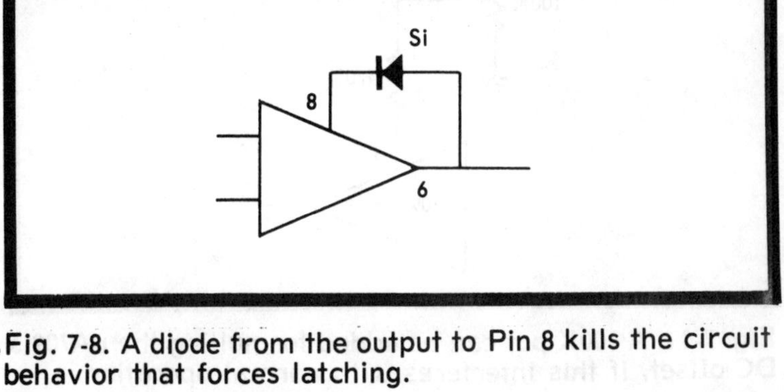

Fig. 7-8. A diode from the output to Pin 8 kills the circuit behavior that forces latching.

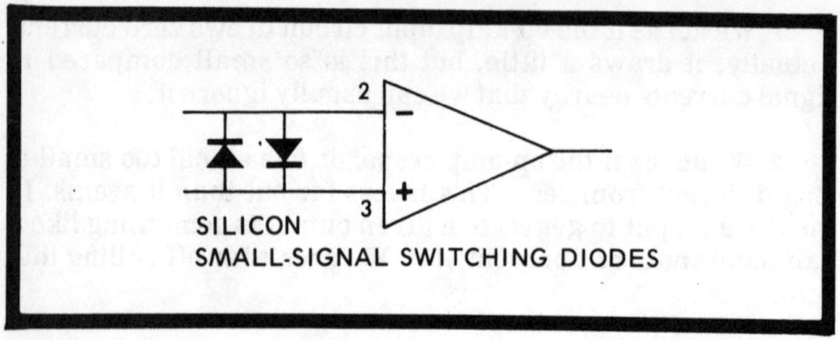

Fig. 7-9. Excessive differential overdrive is prevented by adding these two diodes. They don't affect the 709's typically below-one-millivolt differential input.

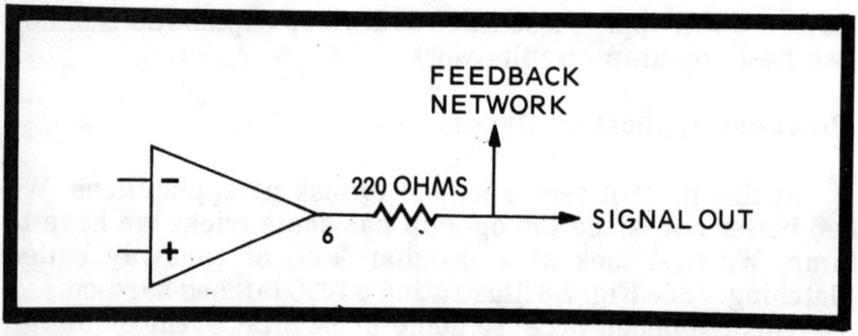

Fig. 7-10. The uA709 can stand a shorted output for about 5 seconds. If this resistor is added, the 709 will survive indefinitely.

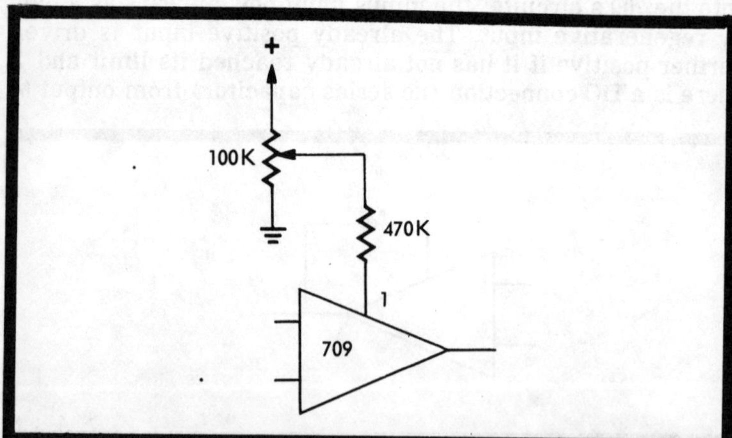

Fig. 7-11. A 100K pot can be added for nulling the uA709's DC offset, if this interferes with normal operation.

input this progresses to a locked state that persists until everything is turned off or the minus input transistor is somehow brought out of saturation. If one of your op-amp circuits stops working, with its output terminal extremely positive, suspect latching. This can be prevented by the protective diode shown in Fig. 7-8, which stops circuit action extreme enough to cause latching.

Excessive voltage across the differential input will force one input transistor into zener breakdown, damaging it. A pair of silicon or zener diodes, applied as in Fig. 7-9, will prevent the differential input from exceeding the 709's rated 5 volts. A 709C can stand a shorted output about 5 seconds. After that, replace it. A 200-ohm resistor inside the feedback loop, as shown in Fig. 7-10, is complete protection. This won't affect circuit operation in most cases.

Supply voltage reversal is bad for any semiconductor, and the 709C is a large, elaborate one. Fairchild's literature suggests their 709's fail in microseconds with wrong-way supply voltages. If there is any chance of this (unskilled op-amp users, power supply transients, etc.) a safety diode in each supply lead drops voltages equally and stops wrong-way current.

Nulling

Unavoidable imperfections in op-amp circuits result in some DC output with zero DC input. This is called "off-set," described as a DC input signal that would produce the observed unbalance in a perfect op-amp having the same gain as the real one. Such a theoretical input can be balanced or nulled out by connections to the input terminals, but the 709 design provides another way to cancel offset. See Fig. 7-11. The 100K pot is adjusted for zero output with zero input, closed-loop.

Frequency Compensation

Op-amp circuits are stable or unstable according to perfectly definite and measurable circuit connections and component values. To those without appropriate background, op-amp behavior seems mysterious or extremely perverse; the expert makes tiny changes here or there while the less knowledgeable amateur looks on and wonders how this magic works. The engineer is applying theory not discussed here; its effect is usually that of frequency response adjustment, also called "compensation."

The engineering approach can be pared to a rule of thumb: Feedback problems can often by cured simply by

CLOSED-LOOP GAIN	RECOMMENDED COMPENSATION
60 db	C1 10p, R1 0, C2 3p
40 db	C1 100p, R1 1.5K, C2 3p
20 db	C1 500p, R1 1.5K, C2 20p
0 db	C1 5n, R1 1.5K, C2 200p

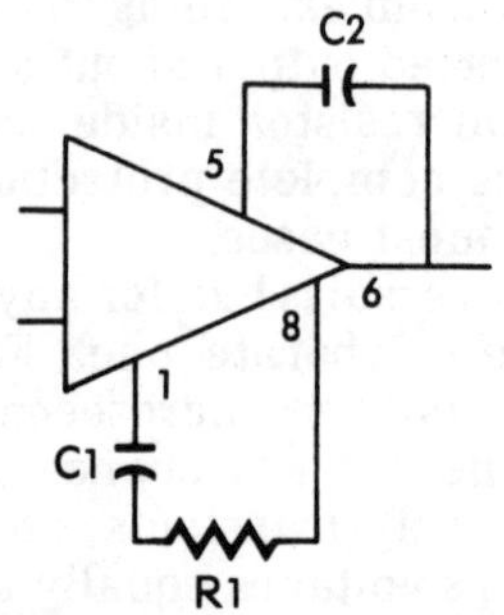

Fig. 7-12. The uA709's frequency gain characteristics can be controlled by adding one of these sets of RC elements recommended by Fairchild. This is an important detail.

reducing amplifier bandwidth. In other words, trouble is associated with high frequencies and if we reduce high-frequency response the circuit becomes easier to live with. We reduce amplifier bandwidth by adding C1, R1, and R2, as shown in Fig. 7-12. The 709C needs at least compensation no. 1 for any application. It's good to begin a breadboard with compensation no. 4, the complete treatment. When things are working right this can be reduced to see if performance gets better or worse (if the application needs fast response or the high-frequency performance).

It isn't exactly compensation, but if the 709C is working into a purely capacitive load, a 50-ohm resistor (approximately) should be added as shown in Fig. 7-13. If this is omitted, the capacitance, seen directly by the 709's output circuits, upsets their workings enough to cause instability.

Inverting Amplifier

The basic op-amp application schematic is the inverting amplifier of Fig. 7-14. Since the signal goes to the minus input,

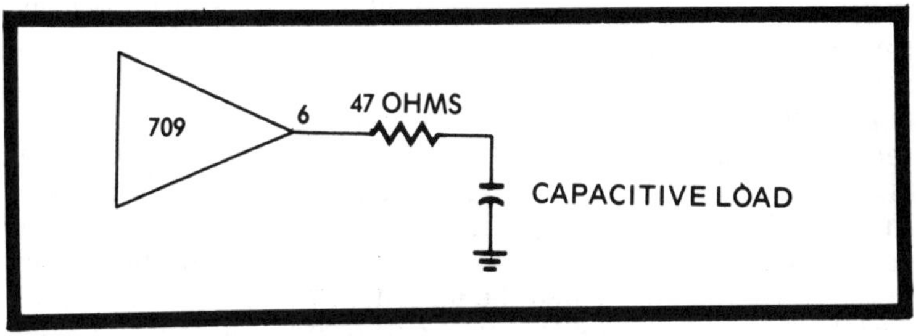

Fig. 7-13. The uA709 doesn't like capacitive loads. A small resistor can be added to stabilize operation.

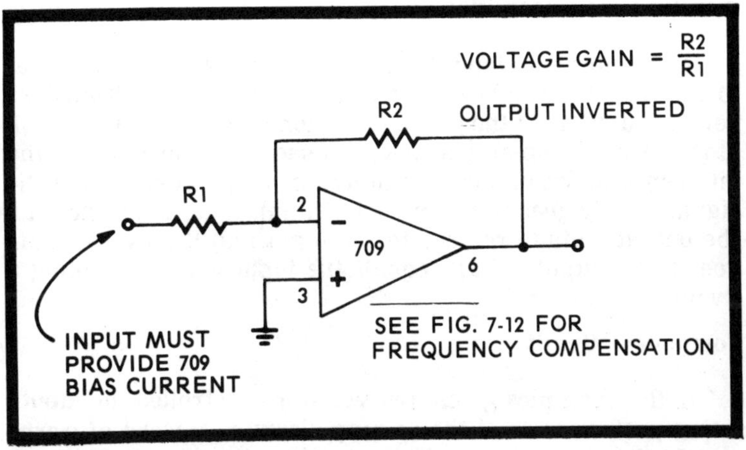

Fig. 7-14. In many applications the input signal goes to the inverting input as shown here.

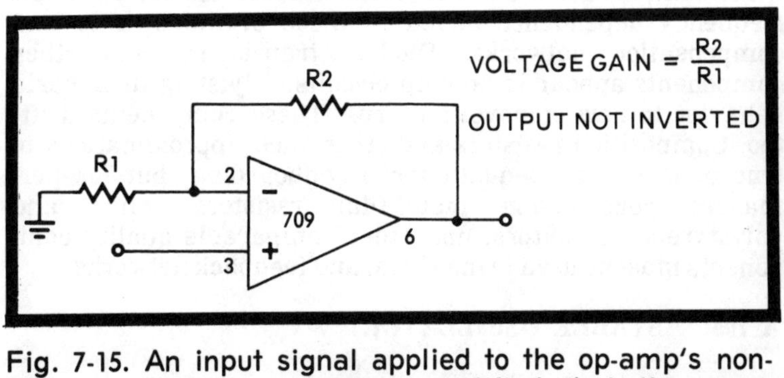

Fig. 7-15. An input signal applied to the op-amp's non-inverting terminal appears amplified, but the same polarity is present at the output. The inverting terminal is still used to fix overall gain.

it appears inverted at the output "inverting amplifier." To find the voltage gain, we remember the input zero-voltage and zero-current rules. Then Kirchoff's node rule tells us the current from source to input terminal equals that from input terminal to output, for a net current equal to zero. And if that is so, then the voltage gain from input to output terminals of the overall circuit must equal the ratio of the feedback to input resistances. This circuit's gain depends upon nothing else, until its value approaches the op-amp's gain. If the feedback resistor is made variable, input-output gain is now variable. For instance, this circuit could convert a stable mercury-cell voltage to a known variable output.

Non-Inverting Amplifier

Be feeding zero input to the minus terminal and our signal to the plus terminal, we get a gain condition controlled as before, but without the phase inversion (Fig. 7-15). If you look at this from the op-amp's view, you see it is responding to the difference between plus and minus inputs, as before, but the signal simply goes to the other terminal. If R1 is omitted, all the output voltage returns to the input and the overall gain goes to 1. Output voltage equals the input voltage, making a "voltage follower."

Component Quality

In the examples given, two resistors controlled the input-output performance of the op-amp circuits. **A word of warning**: If these resistors drift, so will performance. If these resistors have frequency-sensitive characteristics, as wirewound resistors do, their influence will appear as a new frequency dependence added to those of the op-amp and compensation networks. Diodes, transistors, and other components appear in op-amp circuits, adjusting their basic behavior to new or novel patterns. These components drift, too. Composition resistors and other such approximations to true quality are adequate for breadboarding, but for permanent construction metal-film resistors, mica and polystyrene capacitors, and other comparable quality components must be used in the signal and feedback networks.

A 709C ASTABLE OSCILLATOR

This book is like most operational amplifier literature in emphasizing linear and linear-like op-amp applications. But to complete the picture we need to see at least one switching application. There are many of these for op-amps: bistables and monostables, as well as the astable described here. To make a square-wave generator we add three resistors and a capacitor to a 709C op-amp, as shown in Fig. 7-16. The output is a 20-volt p-p square wave with about 0.5 microsecond rise time and a rather remarkable 60 nanosecond fall time. The parts values shown give an output frequency near 1.9 kHz, but the signal is not quite square because at 3.9 megohms, R2 is pushing a practical limit. Maximum output frequency, using a 2.2-megohm resistor and no capacitor at C1 (circuit capacitance and op-amp reaction time), is about 800 kHz. A nice triangle wave near 150 mv p-p is available.

Circuit operation is very simple once we satisfy the key requirements discussed in the next paragraphs. The capacitor (C1) voltage rises from just under zero volts to just above, and down again, as shown in Fig. 7-17. The corners are where the inverting terminal voltage catches up with the non-inverting terminal voltage, which follows the output closely. At the moment the inverting terminal voltage comes within about 600 microvolts (see Fig. 7-6) of the non-inverting terminal voltage, the 709C unlocks, regenerative feedback starts, and its output swings from one to the opposite extreme. Oscillation frequency depends on the size of C1 and the current returning through R2.

The first key point is that the positive feedback loop from the non-inverting input to the output, back through voltage

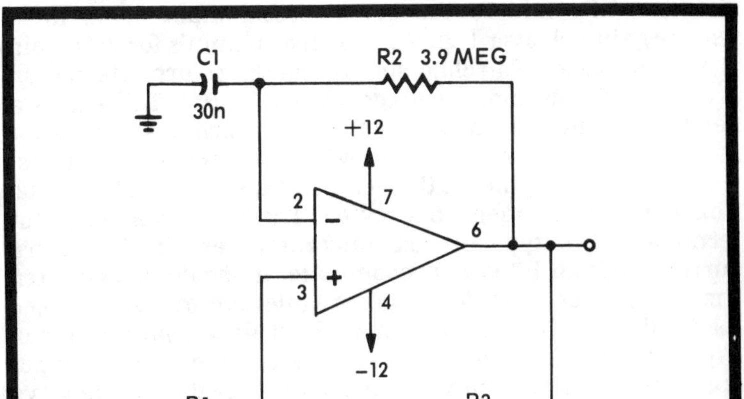

Fig. 7-16. This simple uA709 circuit generates good square waves.

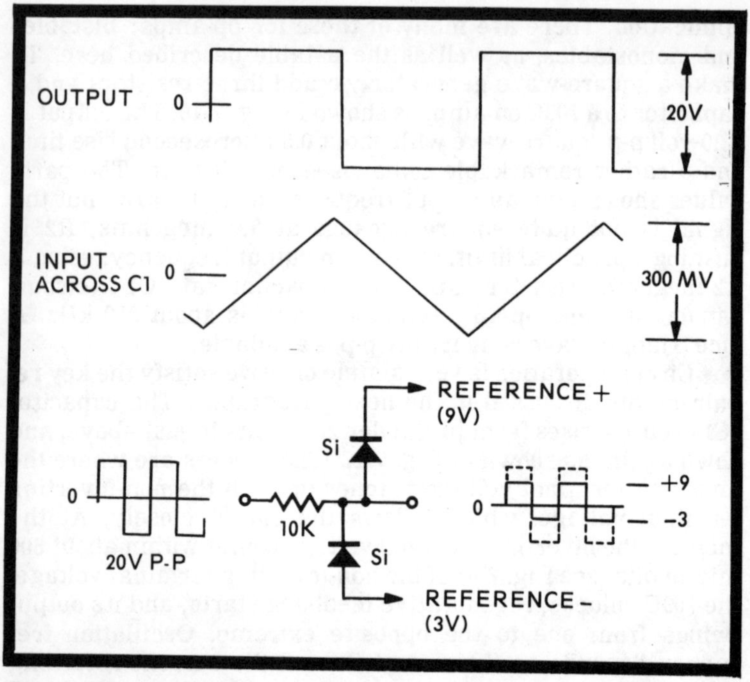

Fig. 7-17. A triangle wave is also available from the op-amp astable.

divider R3-R1 and to the non-inverting input again must feature gain well over 1. In Fig. 7-16 the op-amp's forward gain is at least 15,000 (Fairchild specs) and the return attenuation is about one-thousandth (voltage divider R3-R1). This leaves a net gain around the loop of 15 or better, which is adequate for strong switching action without overdriving the input circuits.

The next key point is the general resistance scale for R2. This is in the low megohms region. The 709C inverting input terminal takes typically 0.5 microamperes; if the return current through R2 is to predominate, it should be about ten times larger, or more. This makes 5 microamps a good choice for feedback—about 2 megohms. The 3.9-megohm resistor in Fig. 7-16 returns about 2.5 microamps and the different input currents for the two output states unbalance the square wave slightly. A smaller R2 is nice, but then C1 must be larger for a given frequency; this is a designer's tradeoff problem.

No frequency compensation appears here because, unlike linear circuit operation, none is wanted. The op-amp cannot go into oscillation during the swing from one state to the other because any beginning oscillation merely helps events finish

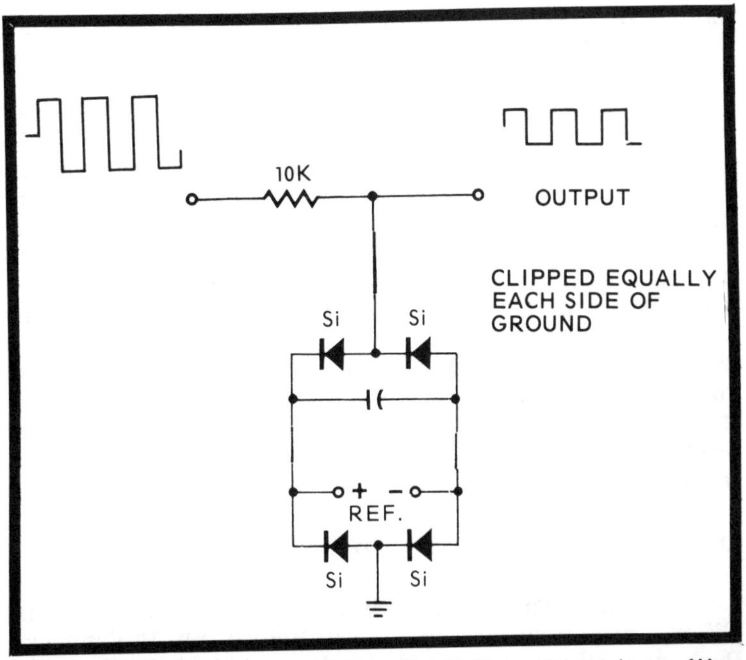

Fig. 7-18. Two ways diodes can limit the op-amp's positive and negative swings.

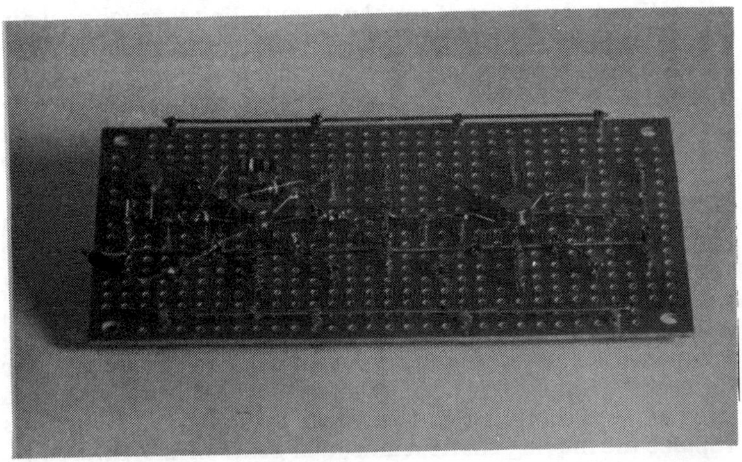

Fig. 7-19. This operational amplifier breadboard carries over simple astable oscillator. Complex circuits can be assembled in its place.

the trip quicker, and then the op-amp is locked and cannot oscillate. Adding frequency compensation merely increases rise and fall times.

Since an output swing of 20 volts is far more than needed for most circuit work, much of this can be discarded for still better rise times. See Fig. 7-18. Series resistor R4 prevents the diode's loading effect from interfering with op-amp operation. As the output rises above the positive diode reference voltage, D1 switches on to prevent a further rise. The minus side clips the same way. The positive and negative reference voltages need not be the same, but 1.5-volt cells will do nicely. The curious (and interesting) wheatstone-bridge limiter can serve using only one bias source, but it cannot use existing supply voltages. If this clipping circuit is to work well, its high-frequency currents must be well bypassed and the diodes must be capable of turning on and off in a few tens of nanoseconds at most. Fig. 7-19 shows the breadboarded op-amp astable.

OP AMPS AND ANALOG COMPUTERS

The amplifier-without-a-circuit, the "operational amplifier," has an interesting history. Before WW II Vannevar Bush and others developed machines called "differential analyzers." These machines solved scientific problems by mechanically modeling the mathematical quantities that represented some physical process. For instance, the acceleration and velocity of a real weight falling through real air are described by simple differential equations, which include corrections for changing air resistance with rising velocity and decreasing height. To make a chart of distance fallen over time the differential analyzer is assembled with gears and cams, modeling the math, so that a part of it actually traces out the behavior of the falling weight.

The differential analyzer's mathematics by imitation proved very successful, but its versatility and accuracy were limited by its mechanical nature. It had to be practically reconstructed for each new problem and used many accurate, difficult mechanical parts. In the 1940s new electronic components and techniques provided a way to make a better differential analyzer. The new idea is now familiarly known as an "analog computer." Electrical voltages and currents replace the moving shafts, cams, and gears of the differential analyzer. Electric charge flows through capacitors, resistors, and electronic circuits to replace the motions of cams, screws, and gears in modeling the problem being solved. But if electric circuits are to copy physical events, the circuits must follow the event's laws, not their own inside inclinations. It turned out this wasn't hard to manage, if somebody could build a really high-gain direct-coupled amplifier.

The result was the operational amplifier. "Operational" means the amplifier works in circuits that imitate the operations of mathematics; its special characteristics are determined by the electrical requirements of its job. As we look at the mathematics of a falling weight, a spaceship in orbit, of chemical processes or the pressure and pulse rate of blood flowing in a circulatory system, we see very simply why electronics principles don't show outside an operational amplifier's terminals. Its input and output signals simply haven't any connection at all with electronics. It uses electronics principles, but its purpose is something else entirely.

For a time after WW II, analog computers were about as popular in scientific work as digital computers. Both did the same job, but the operational amplifiers in analog computers were simpler to use than the shift registers, logic units and complex programming of digital computers. And programming early computers without modern Fortran and other programmer-oriented languages, perhaps working in machine language, was really rough. Very few people had time for both their own work and for mastering the difficult principles of the then new field of computer science.

By contrast, analog computers were simple and easily served as super slide rules for scientists and engineers. In fact, they still do. But as time passed digital computers improved, computer science evolved, and finally digital computers won the contest. The analog computers' inherent accuracy limits and their relative inflexibility (they cannot do electric logic) prevent their seeing wider use. And a digital computer can chew out any degree of accuracy, given enough time for the work. But the analog computers brought us a new electronic component. The operational amplifier now began turning up inside electronic systems. Operational amplifiers now regulate voltages, combine signals, amplify or reduce voltages, and serve many other purposes without putting designers to the work of designing new circuits for these old, familiar jobs. The decade amplifier described earlier is really a kind of operational amplifier, which is why its gain is fixed by outside resistors rather than some inside adjustment. The result is a circuit seen as an electronic component of unusual convenience and utility.

In 1964 designers were often using operational amplifiers even though they were rather large and expensive. Then Fairchild began a design and marketing experiment that brought out their uA709 IC operational amplifier in November of 1965. The early 709s had plenty of new-design engineering and production woes, but they survived as a viable product. A year later Fairchild knew they had a market ten times larger than they had looked for. Raytheon, Motorola and Texas Instruments soon jumped in with their own versions (having names resembling "709") meeting the same electrical specifications. Probably these competing op-amps were all different inside. Early IC operational amplifiers were priced at $50, but by 1967 suppliers were selling commercial grade units at $5. At this writing surplus operational amplifiers seem about to break the one-dollar price level.

An operational amplifier's utility is limited only by the user's knowledge and creativeness. The ones on the market can be assembled into perfectly good general-purpose analog computers. They can perform many jobs in electronic circuits, replacing complex amplifiers specially designed for the job. The operational amplifier is as much a circuit component today as the resistor, capacitor, inductor, and transistor.

Chapter 8

Communications Devices

At one time, radio-electronics circuits dominated nearly all of the electronics field, but recently the subject has diminished in importance, partly because proliferating commercial radio services have reduced the remaining available low-frequency spectrum space nearly to zero and partly because radio-electronics is but a small part of modern electronics. Great adventure is coming, but at this writing it is deferred to the uncertain future. Opportunities include communication with settlements on the moon and with exploring parties living on Mars. In 1970, the year before this book's publication, a small ship was scheduled to land on Mars and seek life there. This experiment might be carried out before 1980 now, maybe. Adventures in space communication must wait. Perhaps when the Russians have established an engineering colony on the moon some of their people will have time to open up space communications for us.

Yet, radio-electronics circuits deserve attention. Understanding how high frequencies and linear circuits work jointly to develop terrific sensitivity to wrong layouts is a real asset for the amateur designer. The man who thinks out the current loops in his small radio receiver will have few problems with spurious pickup and triggering in fast logic circuits. There is also a certain historical interest. Much of radio's history is still on the air in low-frequency services such as loran and aircraft navigation systems. And radio's future is here too, in advanced lorans, omnirange, radars, satellite communications systems, and other ways of solving the communication problem, easier and better.

We hear satellites by radio, and sometimes there are men in space. Indeed, we hear the universe by radio, learning much about how it works by studying the resulting radio signals. This is radio astronomy, now answering some fascinating and important questions. The circuits described here are related to those used by radio astronomers puzzling over pulsars and listening to the continuing echo of the tremendous Big Bang that originated our universe. Now under study by radio techniques is whether our Galaxy's center is exploding, did explode (there is evidence it did) or will explode. It may be all three are true. We could be living in a fool's paradise, in a far more dangerous and inimical universe than many think. Understanding the details, maybe we can do something to improve our chances. If so, our knowledge will be based upon applied radio technology.

But that is far from this book's purposes. Perhaps you wonder why there are no transmitter circuits here. It is because these apply no new principles and their utility is limited. No simple circuits can originate the interesting signals already on the air, and in today's complex world wrong ones make trouble. Airliner crashes have followed their bad-weather navigation systems responding to wrongly designed garage door openers. The Federal Communications Commission requires licensing of all transmitters, which can be managed easily once you demonstrate adequate understanding of the relevant engineering and legal facts. By the time you manage that, you'll know enough to stay out of trouble. But this is a book about electronics, so you must seek those facts elsewhere. We begin with a reasonable substitute for a regenerative detector. Transistor regenerative detectors are hard to manage and are best avoided. The following circuit uses no special parts and is stable (and simple) enough for construction by beginners in communications circuit work.

A SENSITIVE FIX-TUNED RF DETECTOR

The specialized appearance of many radio receiver circuits suggests that they work according to their own private rules. Fig. 8-1 is a fix-tuned detector circuit, which is easily assembled from inexpensive parts to disprove this impression. With trivial changes it could work as an audio amplifier or an AC voltmeter, but it is meant for radio lab work and reception.

Basically, it is a 3-transistor amplifier feeding a diode pump RF rectifier, which serves here as a detector. A DC output terminal is included more as a reminder that a DC signal is available than as any key circuit part. This terminal is useful for some test work, using a sensitive DC voltmeter, and it is interesting to observe response as a signal generator is tuned across the amplifier's passband. In discussing how this circuit works, we can pass over emitter-follower Q3, which is a low-resistance voltage source for the diode detector circuit.

From a signal perspective, Q1 and Q2 are common-emitter amplifiers. For now we ignore the L1-C3 network from

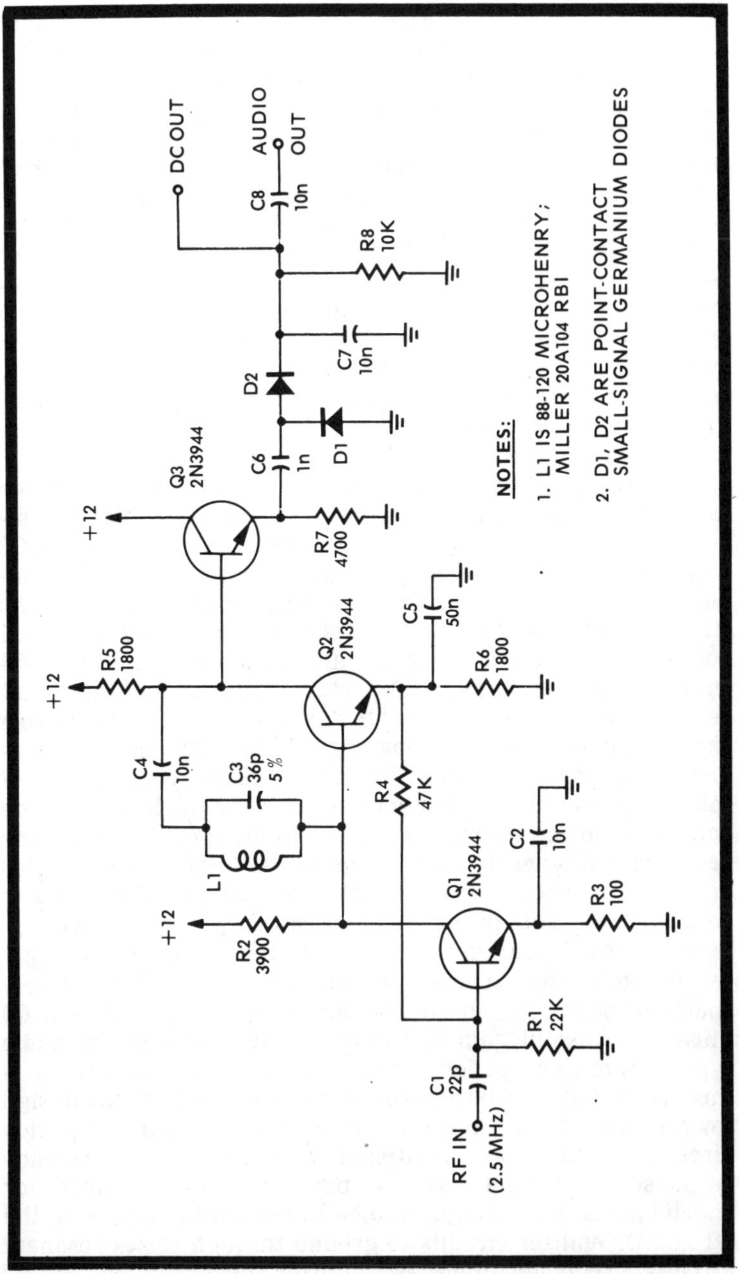

Fig. 8-1. Schematic of the fix-tuned amplifier and detector circuit.

the Q2 collector to base. Signal enters the Q1 base and leaves its collector circuit to enter the Q2 base. From the Q2 collector it proceeds to the Q3 base. Simple. But what is R4? Why are the customary coupling capacitors, bias resistors, and perhaps transformers all left out of the Q1 and Q2 base circuits? Simply to save parts. This design removes two capacitors and four resistors at no performance cost. Q2 is held at signal ground by C5, but its DC voltage follows the Q1 collector voltage closely. It is an excellent terminal for returning "news" to the Q1 base about bias voltages all through the system. If the Q2 collector current rises, then the Q2 emitter voltage rises and Q1 gets more base current. The Q1 collector voltage drops, pulling down the Q2 base voltage to reduce the Q2 collector current. This bias stabilization works through R4, which can be adjusted to get appropriate voltages through the circuit.

In the signal view, this is merely a wideband amplifier (sometimes too easy to achieve, using solid-state components). If L1 and C3 are removed and emitter capacitors C2 and C5 replaced with larger value components, this amplifier's performance can be observed at audio frequencies. At RF its performance will be somewhat less, provided unwanted positive or negative feedback are avoided, but the L1-C3 combination becomes very important. This parallel LC circuit returns the Q2 collector signal to the Q2 base. Since this return signal opposes the incoming signal, the Q2 stage gain is greatly reduced—**except** at the L1-C3 resonant frequency. At this frequency L1-C3 appears as a very high impedance compared to the normal Q2 base circuit impedance, so the return signal is nearly zero. Q2 operates at normal gain at this critical frequency, but at greatly reduced gain for all other frequencies. This performance frequency dependence appears at the overall circuit's outside terminals as "tuning."

Without any special transformers or other closely specified parts, this circuit is easy to build. Q1, Q2, and Q3 must show useful gain at the operating frequency so audio types are out except for very low frequencies. Selectivity is moderate, but can be improved by changing circuit design toward something more conventional in radio-frequency circuits. This frees additional terminals for frequency response controlling circuits, but increases the need for special parts. Selectivity could be improved by bypassing the Q1 and Q2 emitter circuits to ground through series-resonant circuits, reducing gain by emitter degeneration at all frequencies except series resonance for the LC combination chosen. For best selectivity from the fewest parts, it's hard to improve upon conventional filter and tuned transformer circuits.

This circuit can be breadboarded very much like an audio or pulse circuit, if the work is done with an eye to unwanted feedback. The breadboard is prepared with a piece of tin can between the supply buses, soldered to one of them, as indicated in Fig. 8-2. Then the other bus is signal-grounded to the sheet by one or two bypass capacitors. This makes a good ground plane for a low return-current impedance, regardless of which supply line is used as a signal return. And if component shielding is needed somewhere (this is likely), an appropriate piece of tin can be soldered to the ground plane works perfectly. Plain soup cans offer a lightweight easy-to-use grade of sheet tin that is much nicer than plumbing shop supplies.

RF breadboard assembly proceeds with parts leads short but hanging in air about the usual way. It is necessary to think what signals appear on what components, since return coupling from output to input wiring will cause instability. Apparently terrific system performance probably indicates regenerative feedback, which isn't wanted here. Degenerative feedback is equally possible but less noticeable. As a signal generator is tuned near (plus or minus 20 percent) the operating frequency, whistles indicate oscillation. The general circuit tuning can be determined with a grid dip meter, either dipping the L1-C3 network or by observing its strong signal.

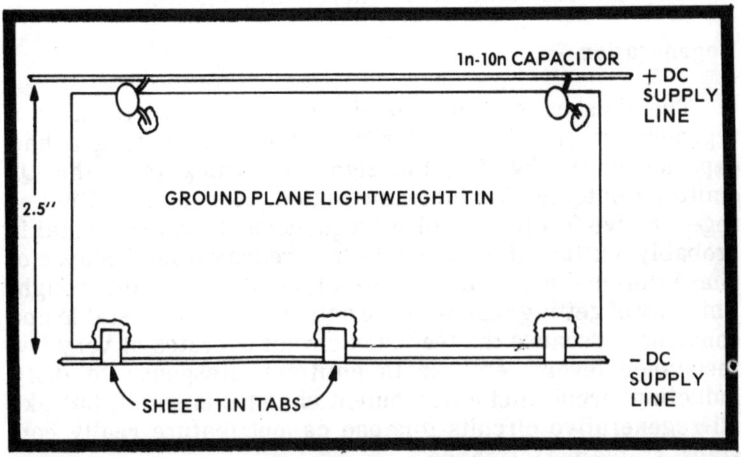

Fig. 8-2. A metal plate can be added between a pair of supply lines to provide a ground surface for RF work. If shielding is needed, it can be soldered to the ground plane.

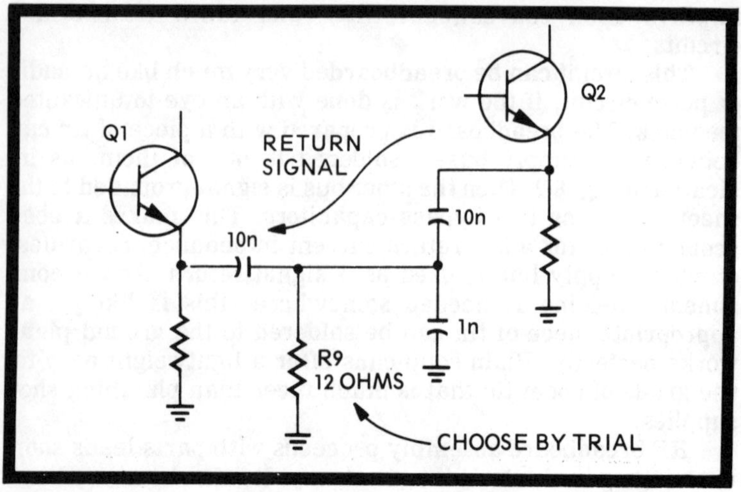

Fig. 8-3. Regenerative circuits are very touchy, but they act better if the feedback loop works at low resistance levels. This circuit is suggested for fairly controllable—but still touchy—regeneration.

This approach quickly gives you a quick-and-dirty breadboard of a utility amplifier without investing in special parts. And it gives you an excellent opportunity to think about RF circuits without getting into something so special its basic principles (which work the same everywhere) are concealed.

Regeneration

The finished breadboard offers a nice way to meet a regenerative communications circuit without having a bad experience. In Fig. 8-3, the signal returning from the Q2 emitter reinforces the incoming signal, so this is positive or regenerative feedback. At RF frequencies the return signal is probably a little out of step with true regeneration because of phase shift in the Q1 collector circuit, but it works well enough. This way of getting regenerative action is relatively stable and convenient because the feedback circuit operates at very low resistance levels (emitter to emitter). Response to drift-inducting circuit and environment changes is small, but like all regenerative circuits this one cannot feature really constant, stable performance.

When the original circuit is rewired in this manner and R9 replaced with a short, the amplifier works as before. But as the short is replaced with increasing resistance, the feedback signal increases. The circuit shows signs of regeneration—sharper tuning, more noise, and eventually, oscillation. The original lab breadboard became highly regenerative using the component values shown in Fig. 8-3. It could be adjusted to the point just under oscillation for maximum sensitivity and selectivity, or barely into oscillation for heterodyne detection of CW signals.

This circuit cannot make a good tunable regenerative detector. Tuning to higher frequencies reduces regeneration or to lower frequencies increases it to excessive levels. An engineer would say loop gain in this circuit is extremely frequency dependent. The reason is that transistor gains are rather strongly frequency dependent at frequencies typically above 100 kHz to 1 MHz. Some really good engineering could probably compensate this somehow, but a really easy engineering solution, developed in 1918, has swept the field. This is Armstrong's "superheterodyne."

Introducing The Superheterodyne

The earliest radio receivers consisted almost entirely of very large antennas. Radio energy intercepted by the antenna went directly to a detector, whose tiny output was applied to earphones in a very quiet room. This was better than a semaphore since it worked at night and in bad weather, and it was valuable for communication across water and bad country. But even the introduction of "tuning" didn't make early radio technology really effective.

Much hard work followed the proof-of-principle in the 1890s, and the first transatlantic communication was achieved in 1901 by Marconi, who did it without amplifiers. Then a curiosity observed by Edison made its way to England, where Fleming came up with his "valve." Later, DeForest made a better valve he called an "audion," but DeForest's vacuum tube didn't work very well. Still, it interested telephone companies trying to transmit speech over longer circuits, so they purchased rights and went to work on improvements. One of the things they did was put a much better vacuum inside. Soon the improved audion was back in radio.

With an audion amplifier, a radio receiving system was no longer tied to getting enough signal from its antenna for a man to hear almost directly. The audion could amplify a signal too weak to be heard, even **before** it was converted to audio, and the audion was also a much better detector than any coherer or crystal. Radio technology developed as the TRF (tuned-radio-frequency) receiver became popular. But the TRF receiver presented engineering problems because its gain and

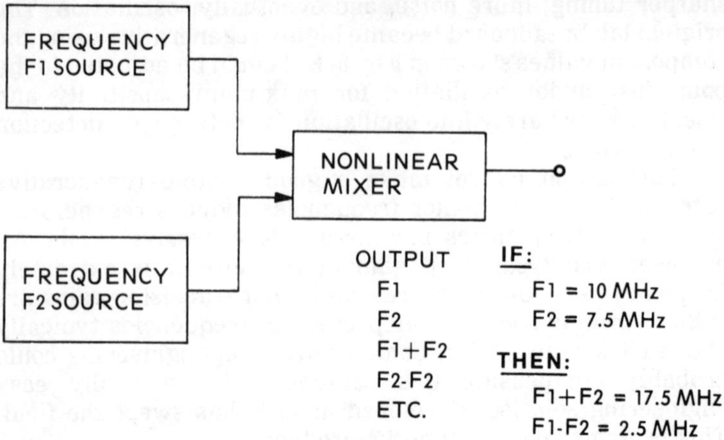

Fig. 8-4. Armstrong's superhet principle applies frequency mixing to get a new, predetermined frequency output from any input frequency.

frequency-selective circuits had to be tuned over the receiver's working range. This nuisance was resolved in 1918 by Edwin Armstrong, who had earlier come up with the regenerative detector. Armstrong built a receiver in two parts: a fix-tuned high-gain section called the "intermediate-frequency amplifier," and a tunable "front-end" that converted the incoming signal to a new, lower frequency for the fix-tuned IF section. He called this a "superheterodyne." Armstrong's superheterodyne concept is basic and will be with us as long as anybody is building communications gear.

Frequency Mixing

If we add (true arithmetical adding) two voltage or current signals in an appropriate circuit, new frequencies appear, as illustrated in Fig. 8-4. These new frequencies carry important characteristics of the original ones such as amplitude, frequency, or pulse modulation. The mixing circuit must be nonlinear in a certain manner resembling the properties of rectifiers. Then the output resulting from this frequency mixing consists of the two original frequencies, a sum frequency, a difference frequency, and sum and difference frequencies of harmonics of both inputs. Superheterodyne designers are interested in either the sum frequency (rarely) or the difference frequency and follow the mixer with circuits designed to de-emphasize all other outputs.

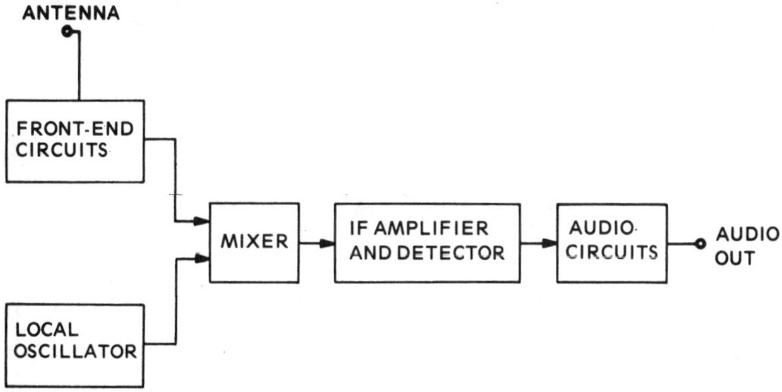

Fig. 8-5. A real superhet has these distinct functional parts. Some of them, particularly the input circuits, may be combined in an inexpensive receiver.

Armstrong applied the idea of frequency mixing to make something like Fig. 8-5. Here, an incoming radio signal is fed through tuning and (optional) amplifying circuits to a mixer, where it meets a locally generated signal. A fix-tuned amplifier, which is really a radio receiver in itself, amplifies the appropriate mixer output and converts it to an audio signal. A superheterodyne is really a kind of receiver within a receiver, with the inside section fixing overall characteristics and the outside part for tuning over a range of frequencies.

A few added circuits convert the amplifier-detector of Fig. 8-1 into a lab-type fixed-frequency superheterodyne. See Fig. 8-6. Left-to-right are the front-end circuit, the local oscillator, and the mixer. Modern receiver design practice is to make the front-end circuits as simple as possible, provided the receiver will do its job. A simple tuned circuit has the basic properties of more elaborate designs so we get by here with L1A and C1, coupled to a short antenna (long antennas complicate the design slightly). Maximum current flows in L1A when the incoming signal is at this circuit's resonant frequency, which is very near 10 MHz, using the recommended coil and capacitor. If there are several incoming signals, the one at resonance generates far stronger currents in L1A than any other. Since a signal appearing in L1B is inductively coupled from L1A, the 10-MHz signal of possible several other signals appears most strongly across L1B.

The other local signal fed to the mixer is generated by oscillator Q1. In this common-base oscillator a tiny bit of collector signal is fed back to the base through voltage divider C5-C6, and since the common-base circuit is non-inverting,

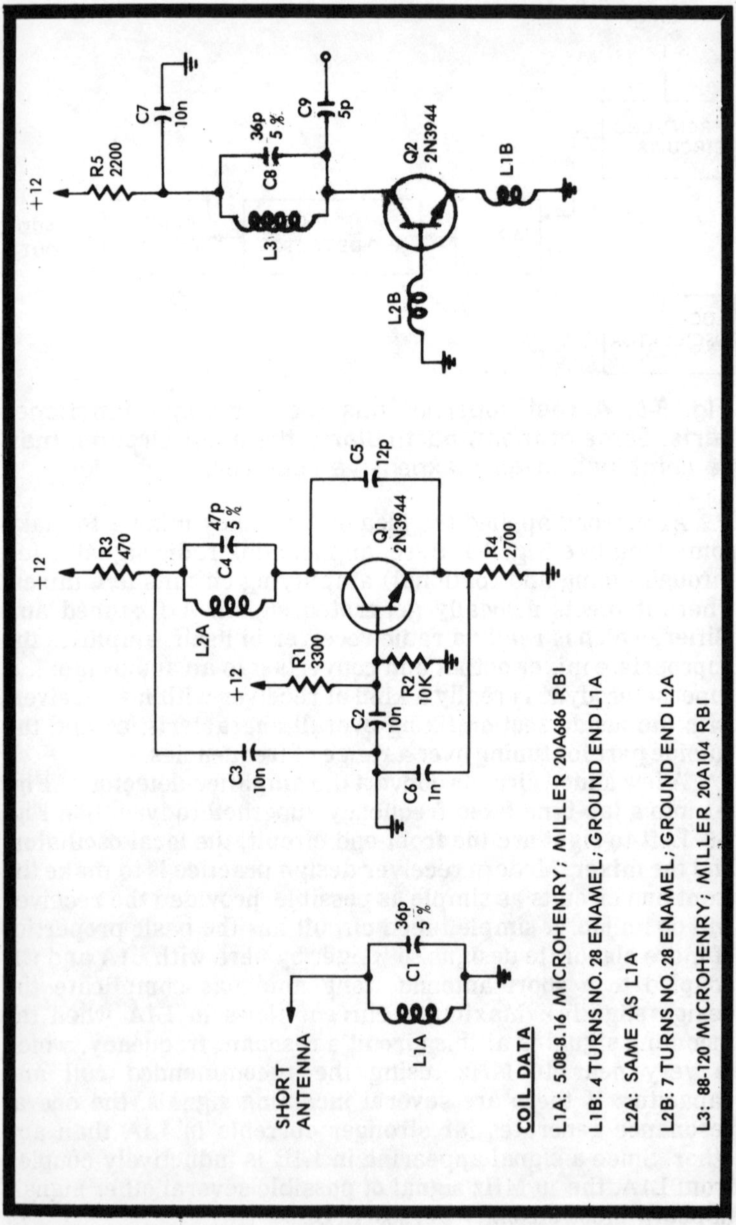

Fig. 8-6. Real front-end circuits for a fix-tuned superhet receiver based upon the fix-tuned amplifier and detector. If you try this before working on more complex circuits, you will understand them better.

this return signal appears as positive feedback. If bypass capacitors C2, C3 and C6 are grounded to the same point on the ground plane, then oscillator currents cannot be coupled into other parts of the circuit. Since there is little leakage to the supply lines through R3 and R4, Q1's only real output is through coupling link L2B to the mixer. These are details, and attention to them now can save you wasted hours later.

The Q1 oscillator signal is fixed at 7.5 MHz by tuned circuit L2A-C4. This is fed to mixer transistor Q2, which rests at cutoff since its base-emitter diode is unbiased in the absence of oscillator input. If the oscillator is working and coupling is sufficient, the Q2 base-emitter diode is forced into conduction every positive half cycle. The local oscillator signal is fed to the Q2 base since the same voltage is needed to push Q2 into conduction whether applied to the base or emitter terminal, but less power is needed at the Q2 base.

Now the base and emitter signals applied to Q2 can mix, generating sum, difference, and other frequencies at the Q2 collector. The purpose of R5 is to decouple the collector circuit from the supply lines and to provide a test point for checking mixer operation. A 2-volt drop across R5 indicates about 1 ma collector current, indicating sufficient local oscillator drive.

With a 10-MHz input frequency mixing with a 7.5-MHz local frequency we can look for 2.5-MHz and 17.5-MHz new frequencies. Actually, there is quite a hash at the Q2 collector with the oscillator signal predominating because there is so much fed into Q2. We add L3-C8 to emphasize the desired output with 2.5 MHz preferred because it is below the input frequency and, so, easier to manage, amplify, and detect. Capacitor C9 is very small to avoid overloading the following transistor circuit with the oscillator signal. The output signal is applied directly to the input of Fig. 8-1.

Construction Notes

This front-end circuitry should be assembled over three inches of breadboard, if hung up on supply lines 2.5 inches apart. A ground plane is needed. In the original breadboard the circuits were too tight and some shielding was needed, but simply a little more room should solve this well.

Tuning is simple, if carried out according to the properties of superhets. After tuning the IF to 2.5 MHz, feed a 10-MHz signal to the input and tune the oscillator to hear it. Then tune the front-end and mixer output circuits for maximum volume. Simple. As I finished this work, using a 4-foot antenna, I picked up WWV. This was in Boston, with WWV at its site in Colorado.

Good—but this receiver is meant for demonstrating principles rather than for shortwave listening. Maybe you can develop it further!

A SIMPLE UTILITY AMPLIFIER

Many radio electronics circuits generate tiny outputs. It wastes time to engineer and breadboard the audio end of every new circuit project. The well-equipped workbench needs something to sub for the proposed audio end while work proceeds on the trickier, more challenging low-level circuits. This can be engineered and constructed, or adapted from some appropriately similar commercial item. That's a good idea: it goes fast and the finished job has long-lasting value.

Some commercial manufacturers plan over-runs on their printed circuit amplifiers and other units, and offer these apart from any consumer hardware. Something like this is just right for a utility amplifier. Mounted in a box, it provides instant audio for new engineering and design work, and for servicing. A test probe can be plugged in for signal tracer application, or other utility circuits can be fitted in or used externally. An example is the simple additive diode mixer in Fig. 4-38. When it is added to the amplifier described here (a foot of mike cable using miniature phone jacks) the result is an extremely sensitive and useful RF signal mixer and test instrument.

We begin with a 360-milliwatt amplifier marketed by Lafayette Radio Electronics. (Anything remotely like it, or something removed from a defunct transistor radio will do as well.) At this writing, Lafayette has offered this and other such units some 10 years or so. It looks like it was originally developed for a tape recorder or a consumer radio receiver. It offers microvolt audio sensitivity, adequate tone, stable performance, and it's a finished unit. It goes nicely into a roughly 2 x 3½ x 6-inch utility box with plastic panel.

The loudspeaker, purchased separately, goes at the right-hand end of the panel. It should be the largest that can be fitted in; these sound much better than really subminiature speakers. A pair of dividers will scratch out a circle of loudspeaker diameter. This marks the limit of a grid of holes drilled to let the sound out. In drilling, small holes serve as centers for finish holes about a quarter inch diameter. Other panel holes provide for an earphone output, high-gain and low-gain inputs, and a gain control (which carries a power switch). A pad of sound-absorbent material placed inside the box improves tone. The amplifier is bolted to the box bottom with clearance on all four sides. Two banana jacks, red and black,

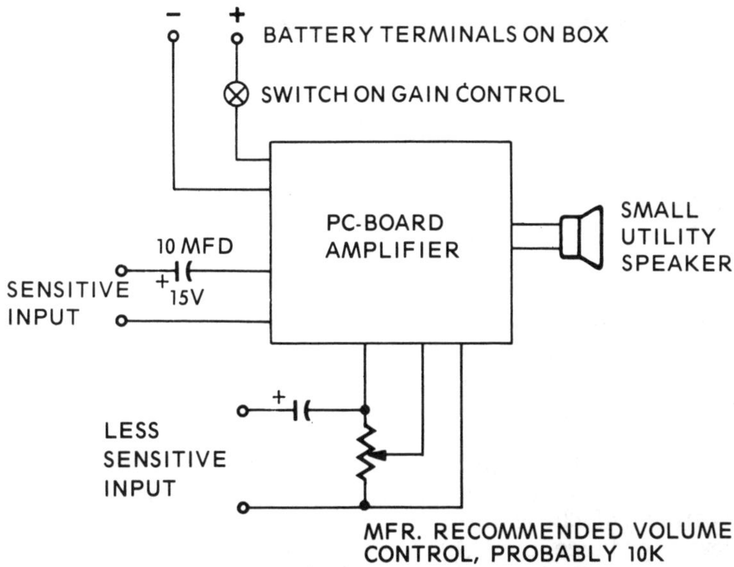

Fig. 8-7. Adding these outside circuits to an inexpensive printed-circuit amplifier makes it into a very useful bench.

provide for DC power input. A protective diode might be a good idea. All other connections are made to the panel, which faces up. This is most convenient for lab work. At this point it's time for wiring. See Fig. 8-7.

It turns out the amplifier's normal input is too sensitive for some lab projects, but fine for others. The sensitive input is brought out at the extreme panel left to a miniature phone jack. The schematic shows an electrolytic capacitor whose positive side is out; this terminal must not see minus or large plus voltages. It is simple to add an appropriate series capacitor where an outside source circuit would violate this rule, and a permanent 47K resistor across the input stabilizes DC voltage there at zero when using a protective capacitor.

A second input hooks into the gain control circuit, where sensitivity is far less than the first input. An electrolytic capacitor is used here too, polarized the same way as the first input.

Placing the batteries outside the case is a good departure from common practice. "Out of sight, out of mind," applies to batteries as well as anything else, and this keeps the batteries away from the electronics too. And there is a certain economy to using the same pencells for all work, rather than having many of them sitting around working rarely.

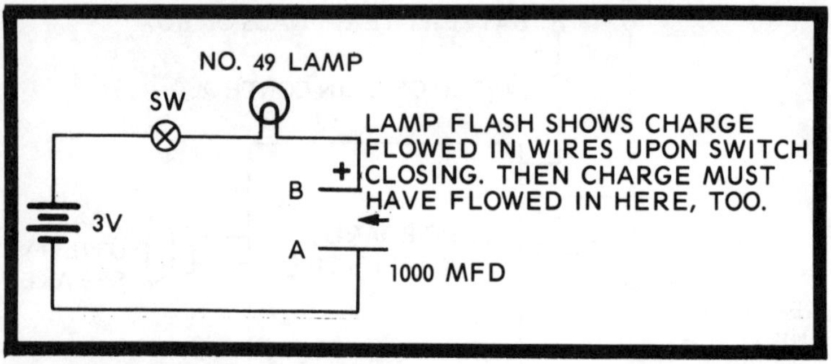

Fig. 8-8. Displacement current is a charge movement through space required by Kirchoff's node rule. Without it a capacitor could not accept or store charge.

Putting an inexpensive amplifier into a box with two inputs and an optional earphone output proved a very good investment of a little time and money. The amplifier's relatively low-impedance input (10,000 ohms) is really about typical for modern solid-state circuits, and there is no need for additional adapter circuits. Every time there is small-signal work to be done, this amplifier comes out to offer instant output. It served in getting initial readings on the whistler amplifier project; it worked through many consecutive radio receiver circuits while doing this book, and it has done well in many other projects.

ANTENNAS AND RADIO COMMUNICATION

Perhaps radio's lasting popularity as hobby and entertainment depends upon its apparent mystery. Somehow "nothing" comes up with sound and speech and pictures; a collection of electronic circuits provides communication with people hundreds or thousands of miles away. There is no tangible connection across this distance, but the signal gets over so there must be one there somehow, and we should know something about it. That's a common-sense thought avoided by many amateurs in radio and electronics. For this dubious convenience they invest a good price in time and money spent on inappropriate radio gear. But the underlying basis of radio communication is too important to ignore—and too simple. To understand how radio communication works, we begin with "displacement current," an unknown that troubled physicists in the early 1800s.

Kirchoff's node rule tells us the sum of currents entering a circuit node is zero. By the 1850s this rule was well established, but it didn't always seem to apply. For instance, a simple experiment showed charge could be moved into a Leyden jar (capacitor). But since it was an open circuit for DC, how could this charge flow in or out?

In Fig. 8-8, when S is closed, charge flows from the battery onto plate A, and from plate B back to the battery. But if charge flows into plate A, Kirchoff's node rule says equal charge must flow out of it; also, the same amount of charge must flow into plate B. How does this happen, since the insulating properties of good glass are well established? In the 1850s men puzzled over this and other apparent failures of Kirchoff's node rule, which was already too well established to permit these exceptions. By 1873, theoretical physicist James Clark Maxwell had an answer. He concluded (and worked it out in complete mathematical detail) that a changing electric field in space transferred charge in as real a way as its familiar motion through wires (but Maxwell never knew about electronics). When the capacitor of Fig. 8-8 is charging, an electric field develops between its plates. The growth of this field is accompanied by a "displacement current," which fully accounts for the charge entering and leaving its plates. Kirchoff's node rule is satisfied, and Maxwell's equations far more than a hint of something new.

Because, as mathematics often does, Maxwell's carefully developed equations told a larger story than the men of 1873 could clearly see. They said an alternating charge flow in space—an alternating displacement current—could set off through space in the self-sufficient manner of light waves, or what we now know as radio waves. Shortly later (about 1885) Hertz used Maxwell's work as a guide in seeking radio waves,

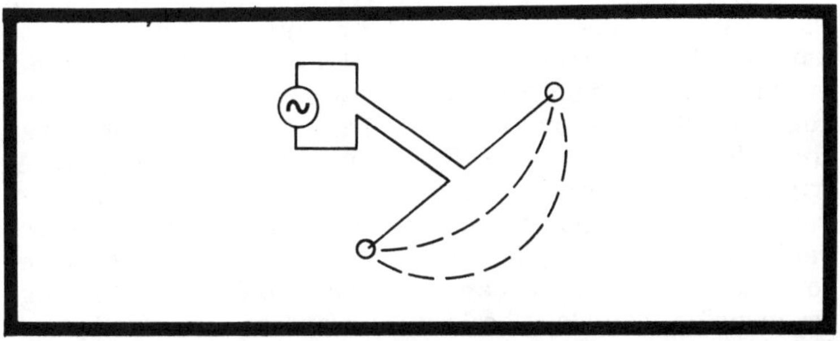

Fig. 8-9. If the capacitor plates are widely separated and redrawn as two charged spheres energized by an alternating current generator, we have a physicist's dipole. This is the theoretical predecessor of the real dipole antenna that works the same way.

and he found them. The discovery was no accident, or any shot-in-the-dark experiment. Pictures of his lab gear show he was deliberately applying the theoretical physics notion of an "oscillating dipole."

Fig. 8-9 shows a dipole (di-pole: two poles) fed by an alternating current generator. Two metal spheres connected to the generator serve as a radio wave source, or an antenna. Each one is an electric pole like those discussed in Chapter 3, but they work as a pair, always oppositely charged with the generator changing the voltage between them continuously and rapidly.

The changing electric field sets up a displacement current between the poles, and some of the resulting combined field breaks off from the dipole to travel freely through space. This is all there is to radio wave radiation. Maxwell's equations, based upon his study of displacement currents and other experimental facts, predict it in complete detail. Hertz looked for it, and he found it. In 1901 Marconi, building upon the work of Maxwell and Hertz, built a system for communicating across the Atlantic without using any interconnecting wires—"wireless." Events followed inevitably and rapidly until in 1965 a 10-watt transmitter in the Mariner 5 Mars probe returned facts about space 191 million miles away. And shortly later, men heard something unexpected, a quiet, non-directional hiss detectable only by the most sensitive radio receivers. It seems to be a still-continuing radio reverberation from the explosion that created our Universe.

The TR Link

Hertz originally worked with very short radio waves, but the limits of early electronics technology, together with the nature of Earth's ionosphere (a part of the atmosphere about one hundred miles up that reflects low-frequency radio waves) combined to force commercial and amateur radio work away from the common-sense light-like short wavelengths. At hundred and thousand meter wavelengths plain facts about radio communication are hidden, and many technicians and radio amateurs struggle with an elaborate mythology trying to describe why they can hear each other. As the high frequencies (wavelengths less than one or two meters) are becoming very accessible, many plain facts, once well known, are emerging again.

As an example, let's investigate the connection between transmitting and receiving antennas under ideal but perfectly possible conditions. We discuss the "free-space approximation" (an excellent approximation in free space), where transmission-reception proceeds along a line-of-sight path. See Fig. 8-10. Let us suppose at one meter wavelength our antenna is radiating one watt, which is distributed equally in all directions. (This non-directional antenna is a fiction, like equivalent circuits, and used in much the same way. It's called an "isotropic antenna.") If we are at any point one thousand meters away from it with another isotropic antenna, what received signal power will we find? We begin by computing power density at the receiving antenna. It turns out that at 1 kilometer this is 7.86×10^{-8} watts per square meter.

A very meaningful way to describe antenna performance is to specify its capture area, the size of the "hole" it makes in the approaching wave (if it made a hole; it's really rather complex). Radio handbooks like ITT's Reference Data For Radio Engineers carry tables of capture areas for various antennas, and it lists the isotropic antenna's capture area as .08 square wavelengths. Since wavelength is 1 meter, we get 6.3×10^{-9} watts effectively intercepted by the isotrapic antenna and available at its terminals. This received power varies inversely as the square of the distance, so that given a piece of log-log paper we can easily draw something like line A in Fig. 8-11.

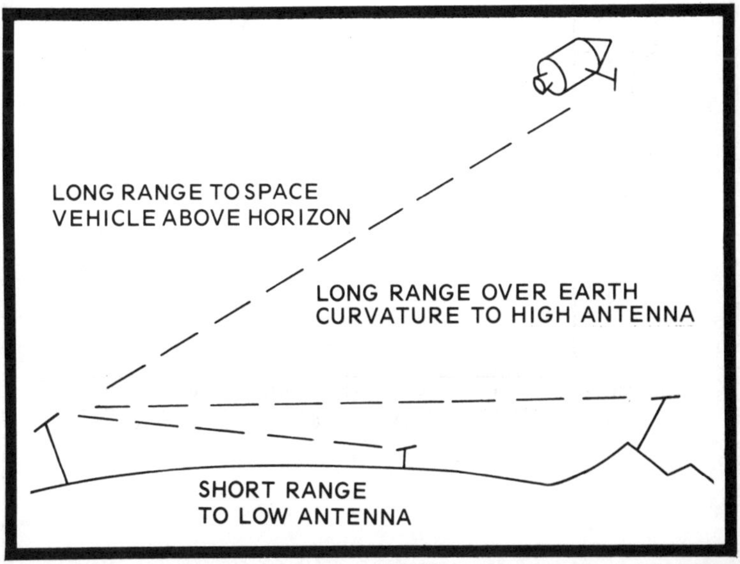

Fig. 8-10. Examples of line-of-sight transmission. In real life, high-frequency signals are often usable outside line-of-sight regions.

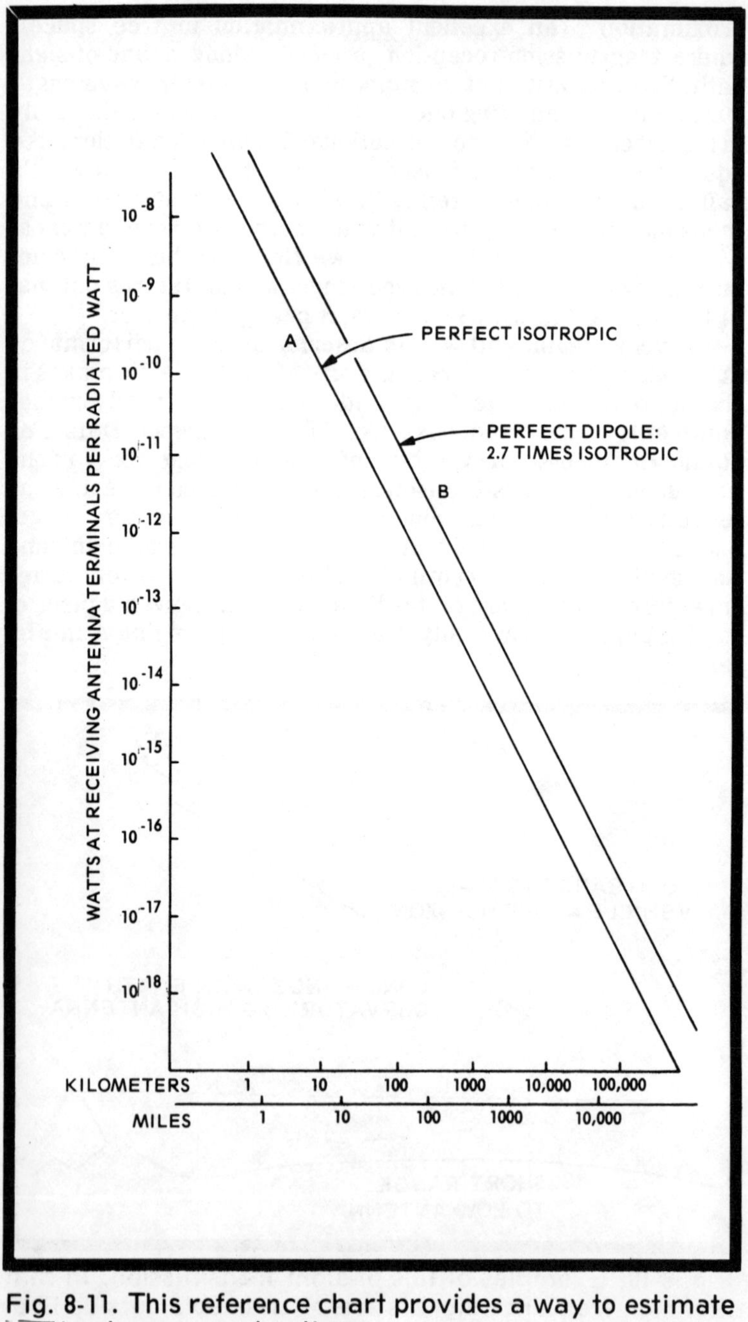

Fig. 8-11. This reference chart provides a way to estimate received power and voltage.

Conversion of this result to something for real antennas is simple. Our plainest real antenna is the dipole (direct descendant from the physicist's oscillating dipole), which is 1.64 times more effective in its best directions than the isotropic antenna. A dipole on each end of the TR link gives us an effective power gain over isotropic of 1.64^2, or about 2.7. This gives us line B, our basic working line for estimating line-of-sight received power for any transmitted power, wavelength, frequency, and range. The correction equation for working out real situations from line B is:

$$Prec = WPt \ \lambda^2 \ Gt \ Gr$$

where:

W is the dipole reference from Fig. 8-11
Pt equals watts radiated
λ represents meters wavelength
Gt is the transmitter antenna gain factor
Gr is the receiver antenna gain factor

Now let us apply this to a simple real case. At the transmitting end we have an estimated 12 watts radiated at 432 MHz from an antenna whose advertised power gain is 7 db. Twenty miles away a dipole is oriented for most effective reception. The path is line-of-sight. How much signal power appears at the dipole's terminals?

At 20 miles the dipole-dipole reference power W equals 1.7×10^{-11} watts. Pt is 12 watts radiated. λ is 0.7 meters wavelength. Gt equals 5 (7 db equals x5 gain factor). Gr equals 1 (0 db equals x1 power gain, or simply no change). Plugging these into the Prec equation, a little quick arithmetic gives us 3.2×10^{-10} watts at the receiving dipole. This point is plotted in Fig. 8-11 to show how it compares with the isotrapic and dipole signal strength lines, and that it is well up in the strong-signal region.

In real life this kind of estimate should be viewed pessimistically. The transmitted power estimate may be high, the receiving dipole may be out of tune or wrongly oriented, there may be signal propagation problems, lossy transmission lines, or poor coupling into the receiver. This doesn't exhaust the list. The experienced electronics worker would prepare for a signal ten to one hundred times weaker than estimated. Yet, this simply computed result is valuable, for it identifies the general available signal strength immediately.

If you're an older reader and all this seems simpler than you expected, the reasons are clear in history. Low-frequency antenna design and communications engineering has always been far more complicated than the free-space approximation, and it always will be. At 80 meters, for instance, a free-space dipole is about 132 feet long placed 1500 feet or more off earth. This just isn't practical. And at these low frequencies, radio waves bounce and skip between earth and ionosphere in a highly irregular and variable manner. There are all sorts of possible explanations regarding what really happens there. This variety hides the essential, basic simplicity of radio communication. As more men go into space and some eventually settle on the moon and elsewhere, the common-sense high-frequency communications technologies should become very important—and interesting.

Chapter 9

"Industrial" Circuits

Only a few years ago, hobby electronics was largely communications electronics. Before World War II, roughly, all electronics was communications electronics except for some very isolated examples in physics laboratories. In those days, electronics hobbyists were called radio amateurs, and they spent a lot of time just talking on the air. That continues in the CB and amateur radio fields, but modern electronics magazines such as Ham Radio, Radio Electronics, and Popular Electronics are emphasizing circuits and ideas that relate closely to modern technology. These magazines are tending to what many call "industrial electronics," and they seem to be getting very good reader response. This extremely important and interesting "other side" of electronics technology uses circuits in ways that resemble the uses of machines a few years ago, and then goes far beyond that. Chapters 4 and 5, on diodes and switching transistors, are industrial electronics oriented. Industrial electronics offers surprises and challenges far beyond anything happening in the amateur radio field. (Developing space communications technology may change this, however.)

The difference between industrial and communications electronics isn't sharp. Most components serve equally well in either field; many integrated circuits do, too. Very often a particular apparatus or type of gear will have some of each in it. A TV set uses communications circuits in its speech section and a combination of communications circuits for picture detail, plus industrial circuits for timing and maintaining the picture tube raster. A radar set uses communications type circuits to transmit and receive industrial type pulse signals. A radar oven employs very simple communications type circuits—to cook food. If you review this book carefully and study the differences, you'll develop some expertise and can be about as unclear about all this as anybody else. Now let's go to a basic, interesting, and very useful industrial electronics circuit—the bistable.

THE BISTABLE

NAND-NOR gates can be assembled into useful systems, but one more element appears in the self-energized logic circuits of control and computing systems. This is the "bistable," a circuit that saves logic signals. A workable bistable can be made from two gates. (It is called a "latch" and hasn't the versatility of the bistable logic unit.) This reduces all of digital logic to one computing element, a quite surprising and remarkable result.

The bistable is a circuit with two stable states. Logic engineers use many terms including "0" and "1," or "set" and "reset" to describe a bistable's state at a particular moment. Since its voltages and currents are stable either way, the bistable remembers indefinitely (or until it's turned off) whatever signal was last fed to it. This enables digital systems to make logical correlations over time, store results, and do extended computations (such as adding two long binary integers digit by digit with appropriate carries). And working in a mode called "toggling," the bistable reduces the input pulse repetition rate by two. The circuit in Fig. 9-1 is easily breadboarded to illustrate the essentials of bistable operation.

The key to this bistable's operation is the DC connection from each transistor's collector to the other's base, along with transistor base connections to a turnoff-polarity supply voltage. Unlike the astable, this circuit cannot relax from a given condition as some capacitor charges or discharges to set up a leap into a new state. To understand how this works we suppose one transistor is in saturation. Q1 is marked in Fig. 9-1; it could as well have been Q2.

If Q1 is in saturation its collector voltage is near ground. The Q1 collector voltage is applied to the Q2 base through the R3-R6 voltage divider, which returns to -3 volts. With the Q1 collector voltage down, the turnoff supply wins, and Q2 is held hard off. But since Q2 is off, there is very little voltage drop across R2. Charge flow down R2 and R4 into the Q1 base is enough to hold Q1 in saturation, even though some flow is lost through R5. Q1 stays in saturation with a very low collector voltage, which is where we came in. The bistable will sit indefinitely with Q1 saturated and Q2 hard off, or with the transistor states interchanged. Except for a catastrophe like the power going off (power failures upset computers badly) only one circuit event can kick the on and off states across the transistors. This is a sharp negative-going pulse applied to Terminal T.

T stands for "toggle." Looking at diodes D1 and D2 we see the Q1 collector-to-base voltage is about the same as the D1 forward voltage so that D1 is almost forward-biased. But D2 sees the Q2 collector voltage as a strong reverse bias. Any signal appearing at T is steered into the base circuit of whichever transistor is on; therefore, diodes D1 and D2 are called "steering diodes." When a sharp negative pulse arrives at the Q1 base, it appears as a turnoff signal, pulling Q1 momentarily out of saturation. The Q1 collector starts going positive because the Q1 collector current is reduced, resulting in a positive-going signal through R3 and C1. This arrives at the Q2 base as a turnon signal, starting a switching cycle like the astable oscillator's operation. The cycle stops when the on state is fully installed at Q2, which stays in saturation until another pulse at the toggle terminal starts another change in state. This is, basically, all there is to bistable operation.

But when we use bistables in logic circuits we must add details. See Fig. 9-2. Q3 and Q4 provide inputs for setting the

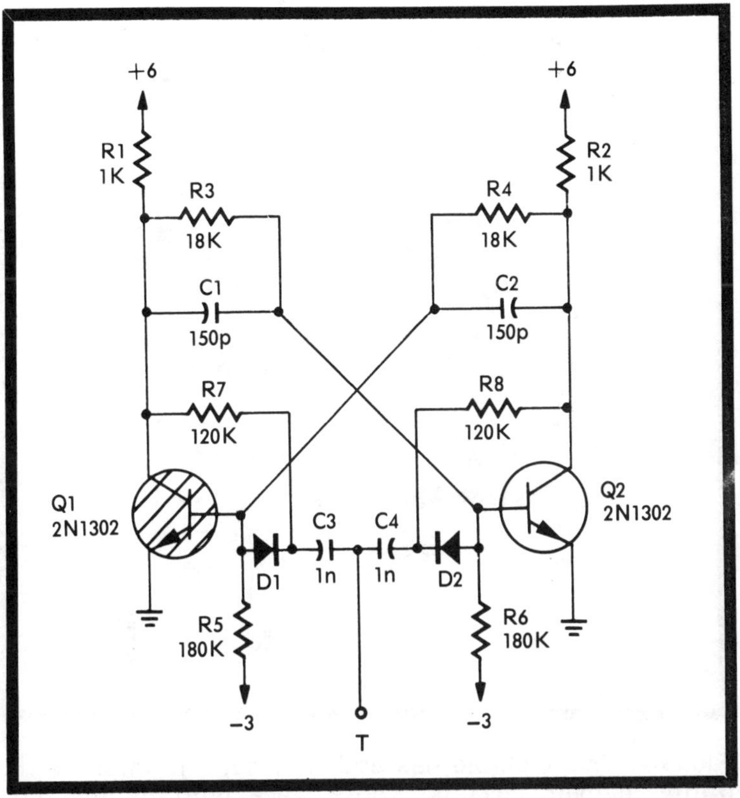

Fig. 9-1. An easily breadboarded bistable circuit. A sharp negative-going pulse at T makes it change state.

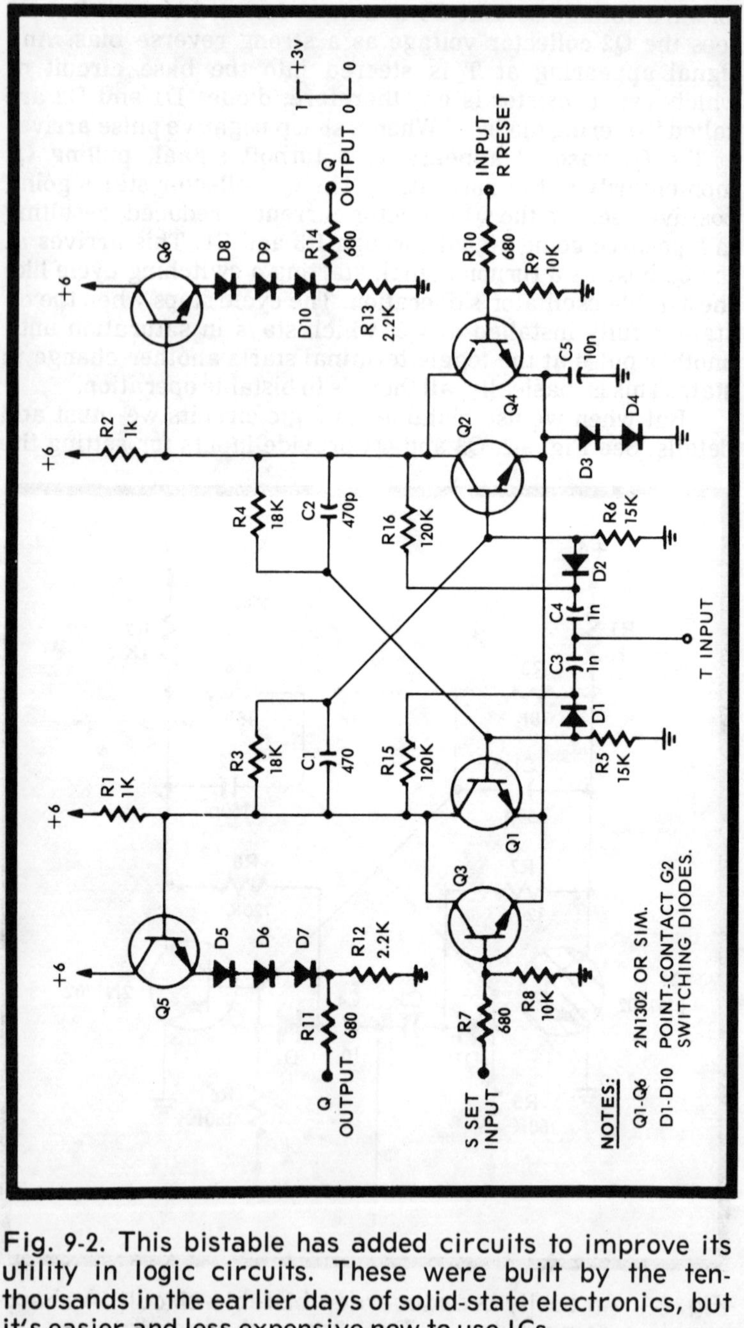

Fig. 9-2. This bistable has added circuits to improve its utility in logic circuits. These were built by the ten-thousands in the earlier days of solid-state electronics, but it's easier and less expensive now to use ICs.

bistable in either state, regardless of previous setting. Q5 and Q6 provide outputs isolated from bistable operation, improving reliability. The minus supply seen in Fig. 9-1 is eliminated by adding a pair of diodes that keep the bistable emitter terminals slightly above ground, so that ground looks like a minus supply to the bistable. And the diodes in the Q5 and Q6 emitter circuits reduce a residual DC level in the logical outputs Q and $\overline{Q}$.

Now we are thinking in positive logic with logical zero at ground and logical 1 at +3 volts or so. If we apply a logical 1 to the set terminal we will find a logical 1 at Q and a logical zero at $\overline{Q}$. If we apply a logical 1 to the reset terminal we find the opposite logical outputs. If we apply a sharp negative-going pulse to T, the logical output states reverse. It's easy to miss a key point here. The set and reset terminals respond to logical 1s to force the bistable to the appropriate state, whatever its previous setting. But the T terminal responds only when a 1 applied to it changes sharply to a zero, and then the bistable reverses states. This detail acquires great importance when we start thinking about how interconnected logic circuits act as a system.

There is a simple way to toggle these bistables without a pulse generator. If a 330K resistor is connected from the plus

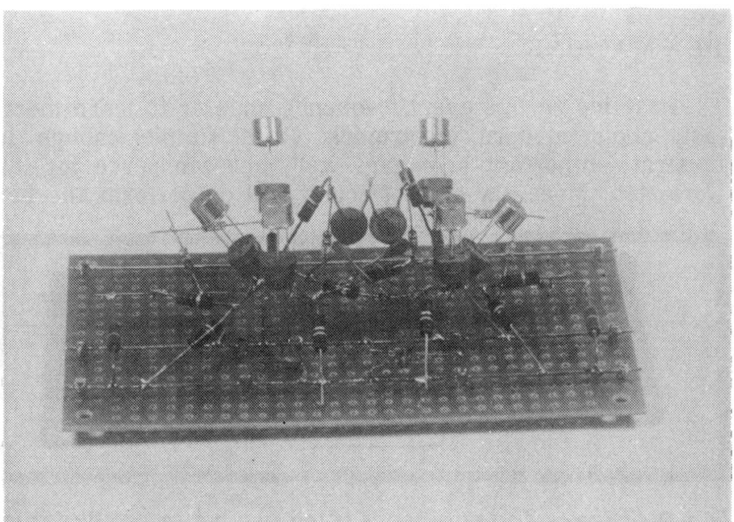

Fig. 9-3. An SRT bistable assembled from discrete components. Compared to ICs, it's huge and works very slowly.

supply line to T, the input sides of capacitors C3 and C4 rest at +6 volts. Now, shorting T to ground generates a negative-going step to toggle the bistable. This method is a little unreliable because tapping one wire against another generates an uncertain pulse compared to that available from a bistable or pulse generator.

How bistables perform logic has a terrific potential for generating confusion at first sight, and the options of positive or negative logic, choices of voltage levels, and various labeling schemes in use, are almost frightening. Just thinking about it all, and practice, will clear things up remarkably. It's a good idea to build one or two of the circuits in Figs. 9-1 and 9-2 (see Fig. 9-3), but these days the logic engineer works almost exclusively with integrated circuits. You will too, once you've experienced their utility and convenience during the following project. A little one or two dollar IC contains an elaborate circuit that works like that of Fig. 9-2, but better.

Fig. 9-4 shows how such a bistable IC is usually described. The designer isn't interested in the inside works, so only the logic terminals (not even the DC supply connections!) appear here. This logical building block is sometimes called a set-reset-toggle, or SRT bistable. At this point we depart almost completely from circuit electronics, and have a try with some logical ICs.

HEX MACHINE

Practice on this event-frequency counter to learn more, fast, about modern electronics. It is simple enough to illustrate important principles well, but can serve for lab work, too. More elaborate circuits will outperform the hex

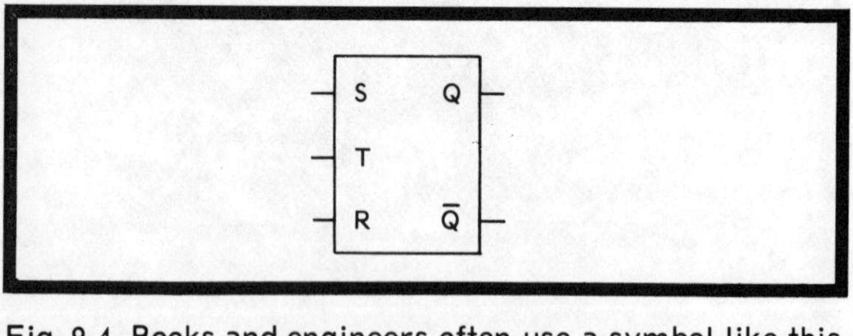

Fig. 9-4. Books and engineers often use a symbol like this to represent a set-reset-toggle bistable. There are many other kinds of counter, computer, and system oriented bistables. Refer to the manufacturer's specs before trying circuit work with real units.

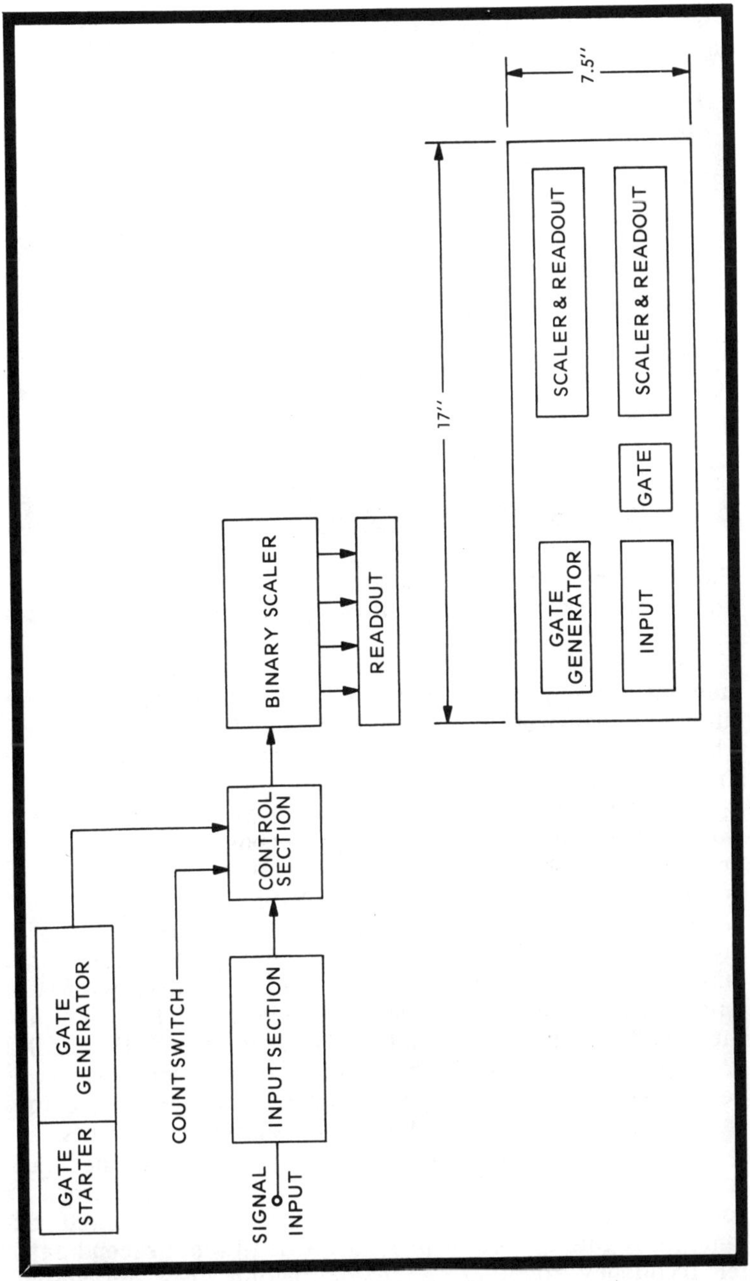

Fig. 9-5. Block diagram and suggested layout for the hex machine.

Fig. 9-6. The hex machine during development. This is a good way to study large circuits and simple systems.

machine (or improve it) but they cost more, are harder to build, and their workings are harder to understand. If you are a beginner, you will find challenge enough here. And if you're a pro, how about trying a **simple** system for a change?

This system can be breadboarded over a weekend, if you work fast, and maybe you'll be happy enough with it to keep it around a while. The key to doing this project well is organization, and we start with an overall view in Fig. 9-5. The block diagram shows the functional parts and the breadboard layout suggests where to put them. The original system was built on a 7.5 x 17-inch plywood breadboard fitted with two pairs of supply lines 2.5 inches apart (Fig. 2-6). This is simply the breadboarding style used everywhere else in this book, grown up a little. Several low-reactance bypass capacitors can be soldered across the supply lines here and there to provide easy return paths for high-frequency signal components.

The hex machine breaks down into five sections. The input section converts the incoming signals into logic pulses. A control section decides if the pulses are to be counted, or ignored. A gate generator provides a fixed tenth-second gate for frequency counting. A scaler section does the actual counting and a readout section, if any is used, provides a visual indication of the number stored in the scaler section.

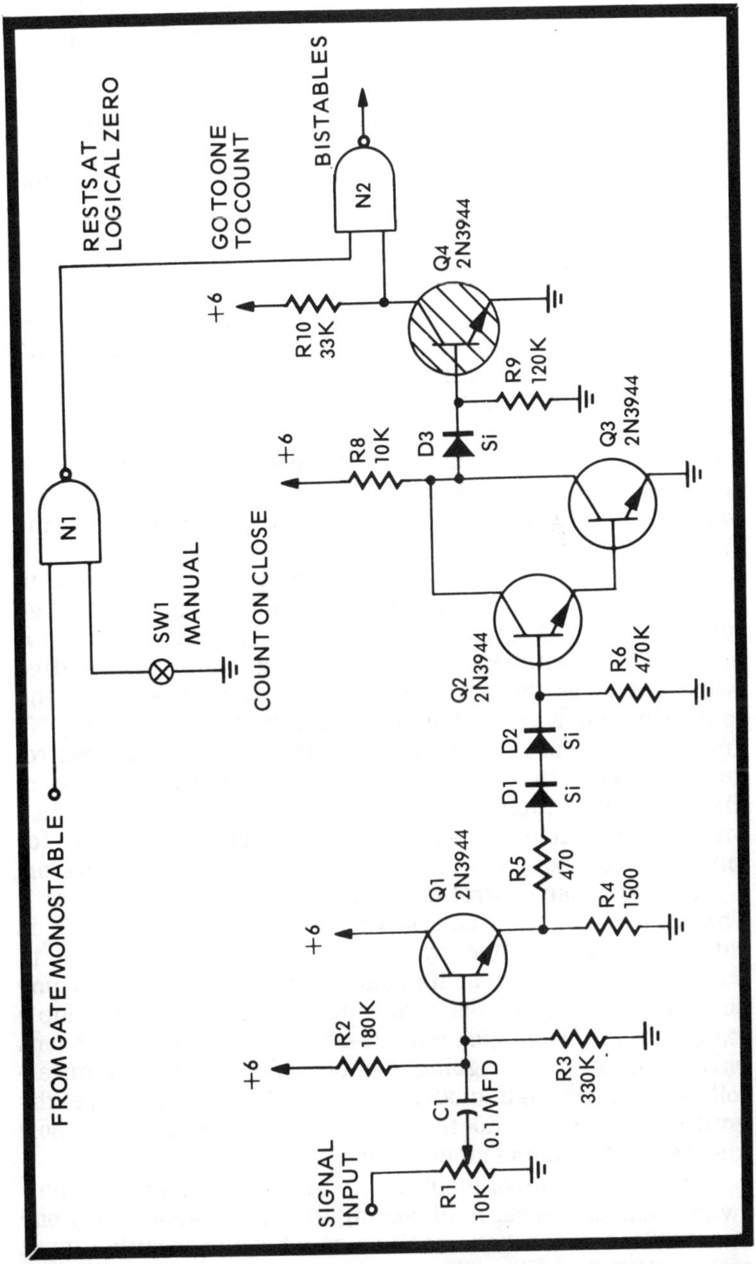

Fig. 9-7. The signal input and count gating sections. All circuits and ICs used in the original model were designed to operate from 6 volts.

It's a mistake to assemble all this, turn it on, and look to see it work. Another approach is better. Build the input and control sections. When these work satisfactorily add the scaler, then the readout, and finally the gate generator. A note about batteries. The pencells that proved adequate for most of the circuit work in this book weren't up to all these ICs and lamps. I suggest heavy-duty carbon-zinc D cells, or alkaline manganese types for this project.

Input And Gate Sections

We begin with a key detail to keep in mind when we review circuit operation. What makes the scaler advance one count? See Fig. 9-7. In the hex machine the scaler advances one count every time the gate N2 output falls from 1 to zero. N2 is a positive-logic NAND gate. It has four inputs that rest at logical 1 if not connected to anything. We leave two floating. One takes the control input, which rests at logical zero except when counting. And the other takes logic pulses developed from the input signal.

This input rests at zero also, switching up to 1s in step with the signal input. If control input N2 is held at logical 1, then every time the N2 signal input switches from logical zero to 1, the N2 output switches from logical 1 to zero and this negative swing advances the scaler one count. Now, when does the control input to N2 rest at logical 1, enabling scaling? If any N1 input is at logical zero, N2 sees the necessary 1 at its control terminal. Since the N1 inputs rest at 1, we can ground any of them to enable counting. Two were used in the original hex machine: one connected to switch S1 to enable counting for any period of time, and one connected to the gate generator.

Now suppose control input N2 has just gone to logical 1. This control signal won't advance the scaler. Q4 rests in saturation as indicated by the diagonal lines drawn in it, holding a logical zero at its collector terminal. Q4 will come out of conduction, allowing its collector terminal to develop a logical 1 if the Darlington pair Q2-Q3 goes on. But this happens only with a positive-going signal arriving from emitter-follower Q1 through diodes D1 and D2. And because the emitter-follower Q1 doesn't invert its input signal, we now know a positive-going input switches Q4 to a logical 1.

Since N2's control input is also at logical 1 its output now swings to a logical zero, advancing the first scaler binary one count. We see the hex machine must count **positive** input swings. Now, we follow the signal back through to N2, picking up details. Q1 is provided to minimize loading on R1 and C1. Notice there is only one series capacitor all through this circuit, an important detail in pulse work but not discussed elsewhere in this book. R5 is simply a current-limiting resistor and diodes D1 and D2 provide a threshold the input signal must overcome, thus sharpening circuit action. Signal levels are adjusted at R1, and R2 should be the smallest resistor that permits Q4 to rest in saturation. This simple circuit works on sine-wave and pulse inputs, but could be replaced with a regenerative trigger circuit. I avoided that here because I think it hides basics. Once you have the hex machine working well, maybe you will want to come up with a better input section.

Gate Monostable

Suppose two positive-logic NAND gates are connected end-to-end as in Fig. 9-8. Unused terminals left floating act like they see logical 1s and we can ignore them. Upon applying power to this arrangement we find one gate output at logical 1 and the other at zero. Each applies the appropriate input to the other. This pair of NAND gates is acting like a bistable and it has the useful properties of a bistable.

First, we will make this logic circuit bistable into a **latch.** See Fig. 9-9. Whichever position the switch is turned to it is forced to a logical zero, with the NAND gate bistable resetting itself if necessary. The latch makes bad switches look like good ones because as soon as either NAND gate input thinks it sees a zero the bistable resets itself to the appropriate conditions. It stays there even though a dirty mechanical switch may produce many zeros and 1s before arriving at last at a steady zero. Logic circuits think even the very best mechanical switches close poorly, so expect to use many latches in logic designs.

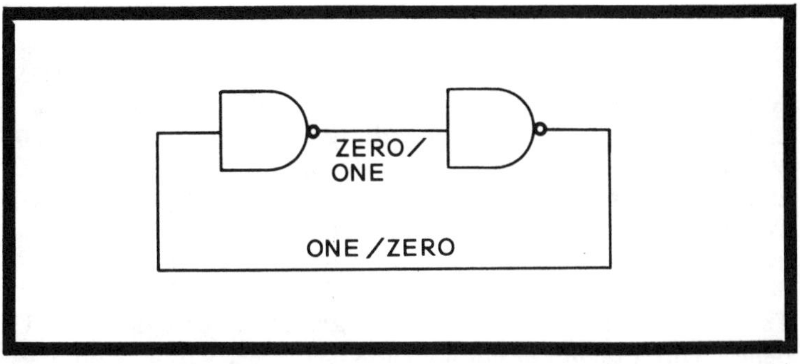

Fig. 9-8. Two NAND gates make a bistable, if wired end-to-end.

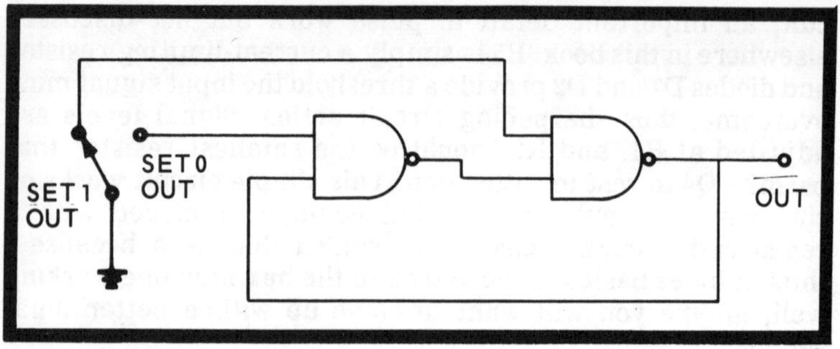

Fig. 9-9. Added control inputs make the NAND bistable into a latch.

The gate generator of Fig. 9-10 is a more elaborate application. Here our NAND gate bistable is interrupted by a transistor circuit whose output rests at a logical 1, but if the bistable switches the transistor circuit can hold a zero output only briefly. Then, as it weakens toward a 1 again, the NAND gates force a reversal to the resting state. This is a latch that cannot stay latched, commonly known as a "monostable." A monostable could be assembled from two transistors, but we use a more complex one here because it has to do a better job than we could expect from a 2-transistor circuit. We want accurate timing and a very sharply gated output. The combination of transistors and ICs does this well.

When the Fig. 9-10 monostable is resting, Darlington pair Q1-Q2 is biased on by the less than 1 microamp base current through R2. One microamp is a very small bias for these transistors, but if Q1, Q2, and Q3 are tested in the suggested check, circuit performance should be adequate. Since Q1-Q2 are on, there is no base current for Q3, whose collector rests at a logical 1. N1 output is a zero and N2 output a 1: this situation can last indefinitely.

Now we turn SW1 from reset to start. The latch N3-N4 develops a negative pulse through C2 that is seen by N1 as a momentary logical zero. Everything reverses around the loop with the N2 output going to zero, Q1-Q2 turned off through C1 and Q3 on because it now gets base current through R3 and D1. The resulting zero at the Q3 collector holds the N1 output at logical 1, even though the triggering pulse through C2 has disappeared. But the logical zero held through C1 cannot last. The tiny current down through R2 gradually brings up the Q1 base terminal from about 3.5 volts to the +1 volt at which Q1-Q2 starts to conduct. The Q3 base bias is weakened; the logical zero at N1 input tends toward a 1, and a regenerative reversal

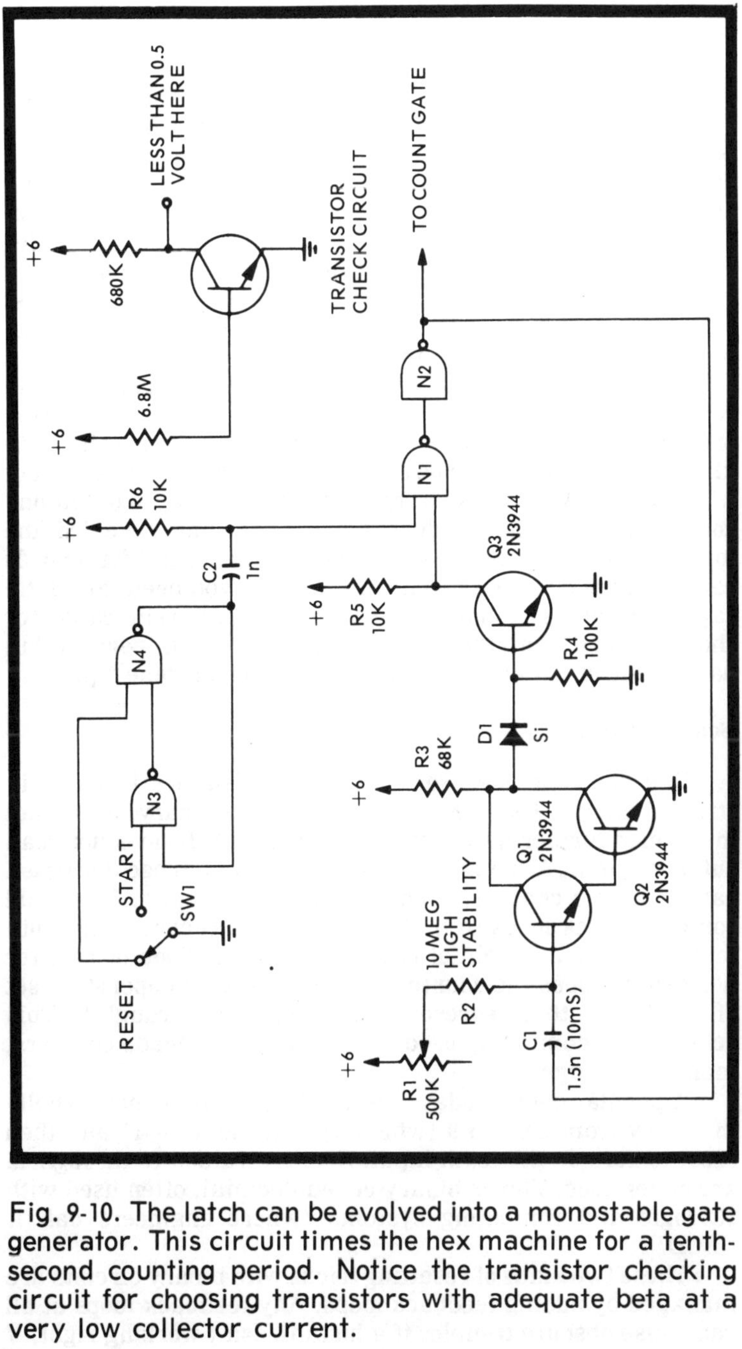

Fig. 9-10. The latch can be evolved into a monostable gate generator. This circuit times the hex machine for a tenth-second counting period. Notice the transistor checking circuit for choosing transistors with adequate beta at a very low collector current.

returns the monostable to its resting state. The count gate ends.

Timing circuit C1-R2 deserves more comment. This circuit is very sensitive to component quality. The transistors require no base current when off, except for a few nanoamperes leakage which we ignore, so the monostable timing depends almost exclusively upon a reliable current through R2 and a high-quality C1. Since very good capacitors are expensive, we keep currents small so we can use a smaller capacitor to generate a given time interval. C1 should be a good quality polystyrene capacitor and R2 one of the best resistors you can find. Don't be afraid of composition types, but use better ones if you can find them.

To calibrate the gate, we feed a known frequency to the counter, such as a 100-kHz signal from a crystal standard (the original hex machine was observed working very well at 200 kHz). Since the gate is to be one-tenth second, timing is correct when the machine runs up from zero to decimal 10,000 in one count interval. Checks during breadboarding revealed the monostable can generate gates shorter than 1 millisecond. It can also gate over one second, but for this you need tantalytic capacitors unless you are a very lucky polystyrene capacitor shopper. The tantalytics can time very long periods but haven't the really good properties of the polystyrene types.

Scaling Circuits

The hex machine uses a string of bistables to count in straight binary. Its output can be read in a binary, octal (not mentioned further here), or in hexadecimal. It does not read out in decimal. A string of four bistables, like those discussed earlier in this chapter, counts up to 15 and resets. If we add some feedback loops so that the 16th count occurs on the 10th input we have a "decimal counter." Then a diode or resistor network decodes the counter's various states to operate a set of ten lamps that represent the digits zero through 9. This scheme was commonly used in vacuum-tube decade counters, many years ago.

A popular choice today is to let the bistables count wholly in binary from zero to 9 (which appears as 8 + 1) and then reset to zero on the tenth input. Counter values 10 through 15 are never used. This is binary-coded-decimal, often used with IC logic and computing systems (where engineers call it "BCD").

These two kinds of resetting tricks with binary circuits are managed by adding feedback loops. Any feedback loops at all can cause obscure troubles if a bistable isn't working right, if

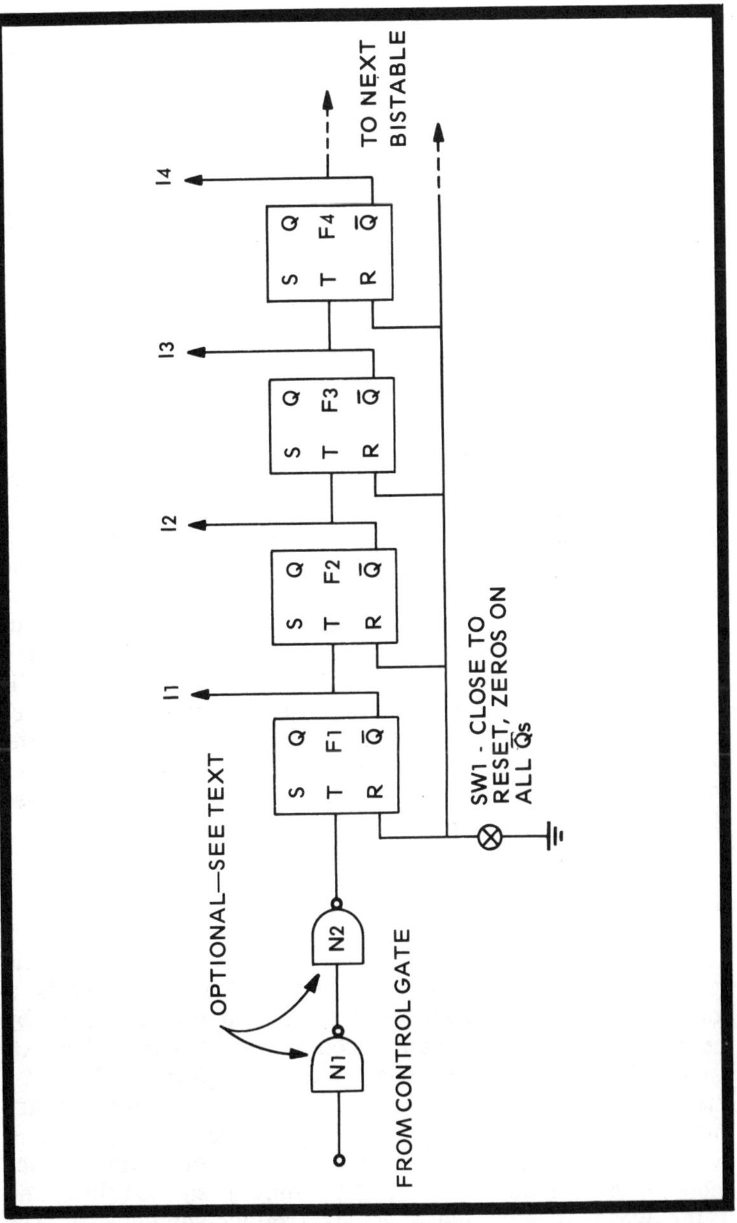

Fig. 9-11. Several positive-logic SRT bistables assembled into a binary scaler. Coupling diodes, resistors, and capacitors are not needed, although the supply lines should be well bypassed.

manufacturer's specs aren't what you think, or if there is a wrong connection. We'll keep the hex machine simple for now. When you understand the principles, you can convert it into a decimal machine—if you really want to.

Fig. 9-11 shows how SRT bistables are strung together to make a counter of any size. You can use true binary SRT ICs, circuits like Fig. 9-1 or Fig. 9-2, or other ICs such as a pair of NAND gates with cross-coupling and steering diodes added. Remember, we are thinking in positive logic, with toggling on negative-going swings. There aren't any outside components shown here because the manufacturer has fitted all that inside the SRT bistables so that we need only pieces of wire.

Switch SW1 resets the bistables on a logical zero. Notice these SRTs are unlike that in Fig. 9-2, which goes to reset upon getting a 1. A pulse entering through gates N1 and N2 switches F1 to its set position **upon ending**, putting a 1 on $\overline{Q}$. The next pulse puts a zero on F1 $\overline{Q}$, which switches F2 to set, and so on. A binary integer builds up in the string of bistables. By reading the voltages appearing at outputs I1, I2, etc. (I for indicator), we can discover what the number is when the counting is finished.

The original hex machine responded poorly to low input frequencies. The pulse input to the binaries had a fall time of 2 microseconds, too slow for the fast ICs. Adding the NAND gates N1 and N2 (two to preserve logic) sharpened the pulses to the required tenth-microsecond speed that triggered the binaries well. If you are interested only in frequencies above 300 Hz, add a regenerative switch to the input circuits, or use slower bistables; these gates can be omitted.

Reading Out

There are at least four good ways to observe hex machine operation and its accumulated result. The simplest is a voltmeter. The number stored in the bistables can be read by measuring the voltage seen at consecutive I points. The ICs used in the original model had a logical 1 voltage of 4.2 volts, and a logical zero near 0.2 volts. When the binaries are toggling on a high-frequency input, a square wave appears at every I. This can be observed well on an ordinary service scope. If there is no square wave the binary isn't toggling, and if the wave is unstable there is trouble somewhere. The square-wave frequency is reduced by two at each consecutive I terminal.

A resistor network provides a way to watch four binaries at once, using a sensitive service scope or a good lab scope.

See Fig. 9-12. Here, a resistor network feeds current from each of four $\overline{Q}$ terminals to a common resistor, R5. R1, the smallest resistor, provides maximum current from the 8-value binary, and R4, the largest, permits a little current from the 1-value binary. The currents add, developing a small voltage across R5 that is directly proportional to the value stored at that instant in the four bistables. Since they count in steps, we get a staircase wave out having 16 levels (zero at the bottom; 15 at the top). Very interesting.

Finally, we come to the conventional way of reading out binary contents. Each output delivers some current to a high-beta transistor (high beta so it needs little current). If the terminal carries a logical 1 the lamp comes on. This arrangement is sensitive to resistor and lamp tolerances, so if you want all the lamps about the same brightness, you'd better make up a check circuit to choose parts that match well. The really critical ones are the lamp and its series resistor. Probably the lamps are sensitive to these details because they are deliberately operated at below normal ratings.

Binary And Hex Arithmetic

The hex machine can be read out in straight binary or in hexadecimal. We begin with a look at binary arithmetic and its relation to the bistables. Since the bistables count in binary

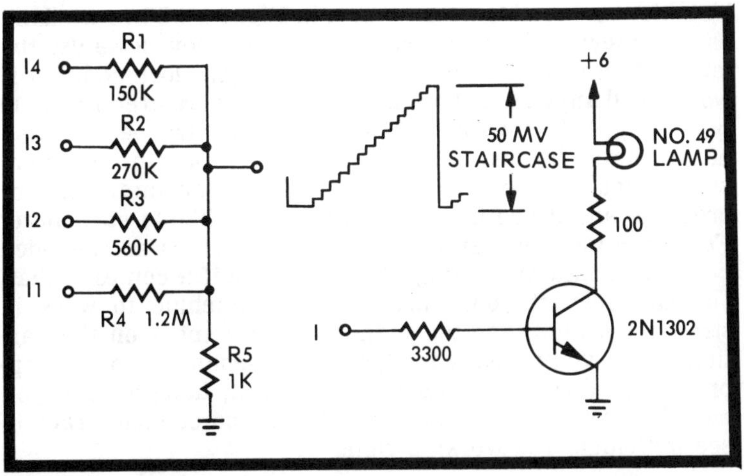

Fig. 9-12. The scaler states can be observed by this simple resistor network, using a sensitive oscilloscope. Bad scaling shows clearly. The lamp circuit can be used to read out bistable states in binary or hexadecimal.

I1	I2	I3	I4	I5	I6	I7	I8...
1	2	4	8	16	32	64	128...

000000 = 0

000100 = 4

000101 = 5

100101 = 37

Fig. 9-13. Table of binary lamp values, and some examples of real readings.

we get an integer by reading, **from the input end,** the states of the individual binaries. Since we typically write our least-value digit at the right-hand end of an integer, you must either feed your binary input to the right-hand end of the string, or swap ends when you write it down, or run the binaries left-to-right and the indicator lamps right-to-left. Supposing you have this settled somehow, you can now work in binary as shown in Fig. 9-13. Binary scalers were used like this for years, and some are still in operation. But with computers becoming very important and many programmers at work, another way of reading straight binary outputs is becoming popular. Take the indicators four at a time, and read them off in hexadecimal.

To read hex you simply take the 16 stable states of the four bistables and read them as shown in Fig. 9-14. Since we don't have numerical digits to represent values above 9 we use the letters A through F. This makes the integers look odd, but it works well anyway and warns you the integers aren't base 10 types. Fig. 9-15 shows how the lamps can be grouped for easy hexadecimal readout. If you have been programming computers, you may even prefer hex to plain decimal, just for practice, and if you haven't, maybe you'd like to start here. For more facts on hex and some useful tables not included here, read data processing literature. And if it comes to that, you can convert your binary or hex machine to work in straight decimal. See the manufacturer's specs on the particular ICs you're using for his recommendations on the appropriate added connections. Don't try it, though, until you have had the hex machine working well and can backtrack to see individual ICs working right.

BURGLAR ALARM SYSTEMS

Teddy Roosevelt: "No man is justified in doing evil on the ground of expediency."

DECIMAL INTEGER:	HUND.	TENS	ONES
HEX INTEGER:	256s	16s	ONES

DECIMAL	HEX
1	1
2	2
3	3
4	4
5	5
6	6
7	7
8	8
9	9
10	A
11	B
12	C
13	D
14	E
15	F

DECIMAL INTEGER	HEX INTEGER
3	3
9	9
13	D
16	10
22	15
30	1E
31	1F
32	20

Fig. 9-14. A quick introduction to hexadecimal. See a book on data processing for details.

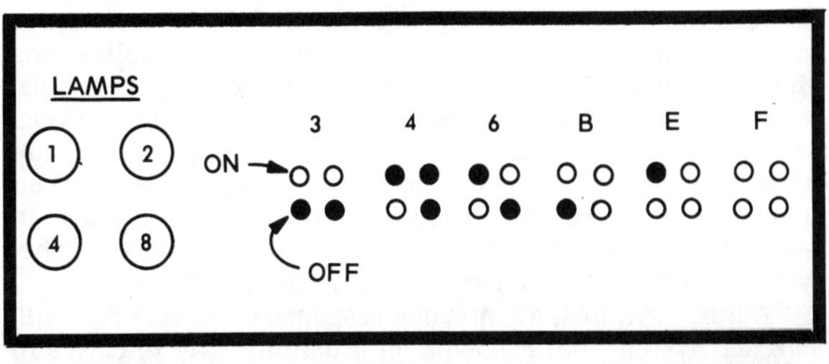

Fig. 9-15. It's easy to read hex from four binary indicator lamps arranged in this pattern.

John Stuart Mill: "We can never be sure that the opinion we are endeavoring to stifle is a false opinion, and if we were sure, stifling it would be an evil still."

Originally, we had planned a brief description of how some industrial electronics concepts and circuits apply to burglar alarm system design at this point. But daily news stories of late 1970 and early 1971 encouraged a further development. Published news stories suggest electronic spying is covered by a moral fog, or at least it all depends upon who is doing it. The essential difference between electronic spying and any other kind escapes me completely, and I offer the following perspective.

The moral principles concerning electronic spying are at least as clear as the technical ones. Any man who watches and listens secretly, who steals private information or who observes, accumulates, and electronically correlates to manufacture it, is a thief like one who covertly takes private property. Further, more or less open attacks and trials under a cloak of justice, using such stolen knowledge, are as vicious and evil as any perpetrated with bludgeon or knife under cover of darkness.

Modern electronics is finding an unfortunate new work in personal, industrial and civic spying. This moral gangrene has festered deep into leading parts of the American system that most need to be free of it. That is all a filthy business and I sincerely hope you will stay free of it.

Intruder Response Systems

The circuit that fires a bell alarm is a nice idea, but the careful engineer begins with a general look at what the problem really is and what is available to solve it. The obvious answers are not parroted here, and instead we look at how basic principles may be turned to these new applications. "Intruder response system" reminds us that any particular detection-and-response system can be worked out several ways. This is pretty much what we have to do since individual circumstances and needs are too variable to provide novel but generally useful designs. Once several possibilities are investigated, the best thoughts can be worked up into a finished system. We begin with a reference to logic circuits.

A burglar alarm or intruder response system is basically a logical system that responds in a definite way to cues supplied by sensors. Usually, the sensors are switches, or subsystems that operate switches. The response may be as simple as firing a bell circuit, or it may include remote indicators, a taped phone call to a specific address, or camera actuation. It may start a defensive system which must not harm the intruder! (A case reported in mid-1970 as "outrage in Iowa" involved an admitted burglar who ran into a booby trap and got hurt. He sued the property owner—and won!) Lights may go on or off, and so on.

Since the system's sensors are switches and switchlike subsystems, all the logic rules of digital circuits apply. See Fig. 9-16. Digital logic is used both for its virtue of stability and reliability, and because a little logic design makes the system act in a manner that is not immediately obvious or predictable to an outsider. Surprises upset intruders—and the unwanted visitor who cannot guess what happens next, or how to stop it, may depart without further adventure.

The switches in Fig. 9-16 can be any of the devices discussed here, or something new you may originate yourself. We see that three switches can be assembled into AND or OR gates, depending upon the logic used, and that control and sensor switches can be combined logically into more elaborate systems. It is important to think about what happens if parts of the system fail in plausible ways such as power off, a bad switch, and wrong operator settings.

Mercury switches and magnetic reed switches are the two most convenient sensors. Pressures switches (mat switches) can be used, but they are more expensive ($4-$15 vs $1-$3).

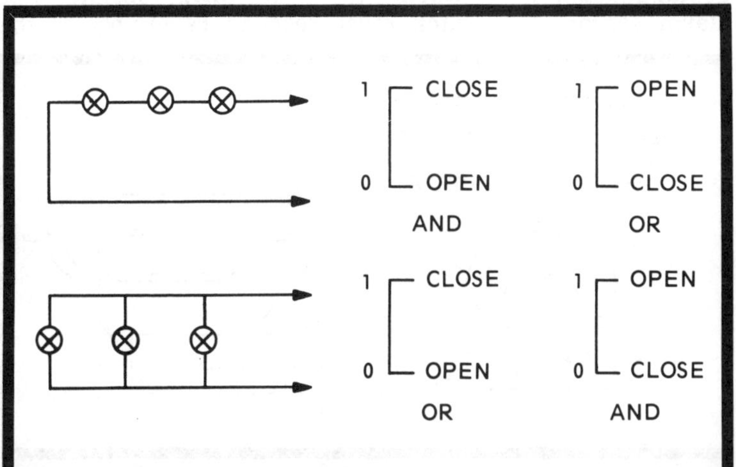

Fig. 9-16. Sensor switches assemble into circuits just like parts of logic systems.

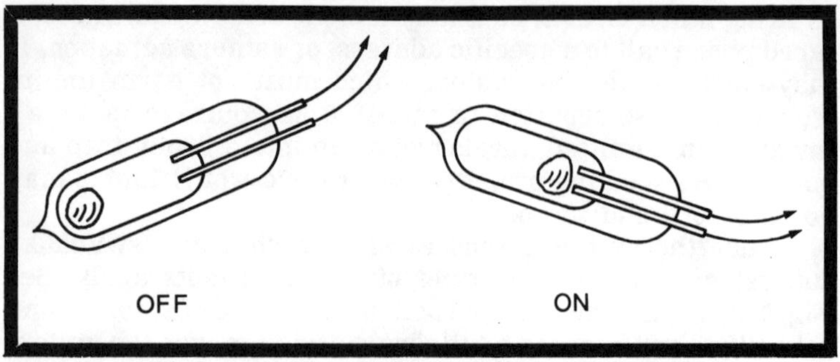

Fig. 9-17. A mercury switch contains a rolling mercury ball. The ball never wears. **Be careful** soldering leads near the glass.

Mercury switches are based on a sealed glass container with a rolling mercury ball that completes a circuit when it rolls down against a pair of terminals. See Fig. 9-17. They are reliable and can switch large powers, but must be installed without strain on their flexible connecting leads. A mercury switch can be operated by opening a hatch or cover (tilt), by a cam action on a door or window, by motion (if mounted almost-on) and by similar means keyed to gravity or acceleration.

Magnetic reed switches are available in large variety. See Fig. 9-18. One type latch closes or opens by magnetizing or demagnetizing its contacts; these are inappropriate. For alarm sensors the switch must respond reversibly to the

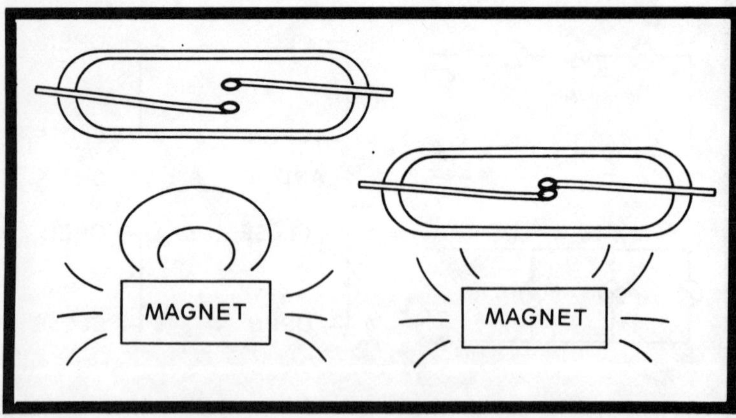

Fig. 9-18. A reed relay switch closes when a magnet approaches it. A stronger magnet works at a greater distance.

motion of a nearby magnet. Single- and double-throw types are available. A key virtue of magnetic reed switches is that installation requires practically no accuracy at all. Inaccurate workmanship or the house settling that makes a door sticky won't upset your alarm system.

Pressure switches are typically pads or strips tending a little over ¼-inch thick. Installation and concealment may be difficult because their sizes make them fairly easy to detect, if the intruder looks for them. Good system design applies several, placed in next-to-most obvious locations under large mats or rugs. Plan against natural wear which will outline the circuits in a mat or rug for all to see!

If you don't usually think of microphones as switches, see Fig. 9-19 for a nice conversion. The amplifier is utility type like those discussed in Chapter 8. Instead of feeding a loudspeaker, it operates a transistor-relay circuit through a diode rectifier. The amplifier's gain control adjusted sensitivity, and a capacitor across the relay confers some immunity to brief external noises.

Apartment owners who live alone may have occasion to wonder if their door has been opened during their absence. The remedy is a non-resettable counter that advances one

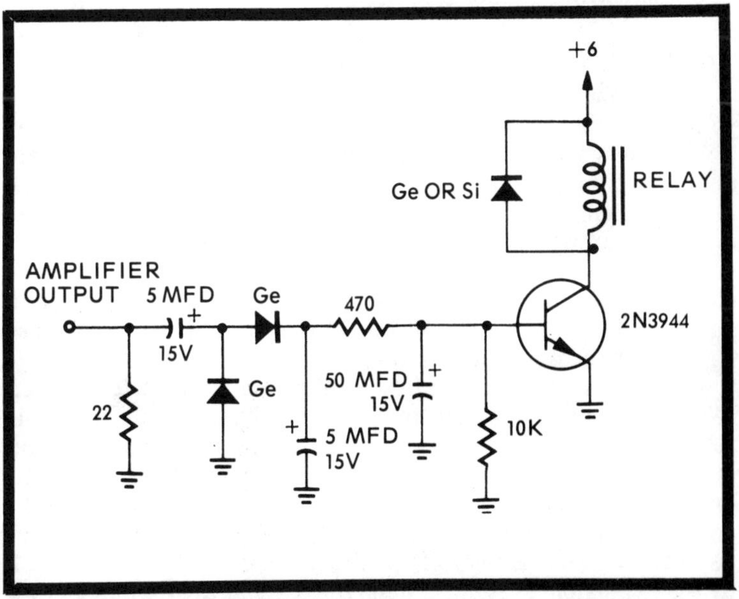

Fig. 9-19. This simple rectifier circuit can be added to a utility printed-circuit amplifier to close a small relay if a microphone picks up excessive noise.

count each time the door is opened. Mechanical and electrical counters are available from surplus and general markets. A mechanical counter must be placed on or near the door. An electric counter (perhaps fired by batteries; it draws current briefly when fired) operated by a magnetic reed or a mercury switch can be located almost anywhere, even outside. The counter should feature at least five decades, a real obstacle to anybody who tries to reset it. Practice will develop your natural ability to tuck away a number and recall it later, so you can be warned that you've had a visitor.

As you engineer your alarm system, think about what it should do if it fires. A loud ringing bell is an obvious choice, but perhaps something less obvious is also more appropriate. Your system could turn on lights, fire a hidden automatic camera, or simply turn on a warning light outside somewhere, or operate a quiet buzzer placed next to your bed (with phone handy in the same room). Whatever the details of your situation, you can work out an appropriate alarm system and you'll probably find it a simpler job than most of the circuits in this book. The engineering principles of breadboarding and system mockup apply as well to intruder response system design as anywhere else. You can finish a very good system if you design and model it well.

Telephone

Telephones interest snoopers because you talk right into yours and you generally have something important to say. Telephone signals are strong and easily copied. An elaborate phone tapping technology has grown up over the past decade. Many unofficial tappers find some unofficial connection right into the company's offices and records, and civic tappers seem able to get active help. But even without assistance, it's simple to tap a phone without upsetting owner, phone company, or circuits. The appropriate technical principles are simpler than much of this book.

Also, it's possible to alter a telephone so that it is an effecitve microphone at all times. There was a small uproar in 1970 (it should have been very much stronger) when a number of State governors discovered very private phones on their desks were converted to permanent mikes, with workmanlike connections through the phone system to an undisclosed central office. A remedy, if you must have a phone, is to invest in an extension system. Then you can simply unplug your phone, and wonder about radio bugs.

Some technicians are making fabulous money in electronic spy gear. For a better perspective on this field try

Spindel's book, "The Ominous Ear." You might lose interest right there, and I hope you do. Of course, if an opportunity comes for you to help the forces of right and justice, why not? Be advised there is a remarkable body of opinion now that there simply aren't any significant forces of right and justice. There are just significant forces. If you can unerringly and convincingly pick one that is entirely in the right, you have just solved a vexing problem that has troubled great men through all recorded history. Appropriate here is a quote from William Thackeray, an English novelist who wrote and thought in the mid 1800s:

"The wicked are wicked no doubt, and they go astray and they fall, and they come by their deserts, but who can tell the mischief which the very virtuous do?"

AUTO STARTERS

The rather peculiar design of most automobiles, engineered largely for the maker's convenience, hides their application of simple and basic principles. But if we think about those principles (which must always apply), maybe we can look inside the machinery without taking it all apart. The central part of a car's electrical system is its battery and starter circuits. Here are two simple devices to give us some idea what's going on in there, if troubles appear. We use the same principles of Thevenin's theorem and Kirchoff's node and loop rules we have seen at work in electronic circuits.

A typical problem is a car that starts hard and we'd like to get some facts before tearing into things. Weak starting might follow corroded cables or bad connections, or a sick battery. Let us suppose reasonable examination has revealed nothing convincing (high technology looks pretty silly uncovering facts anybody can see). We can get pretty far into the circuits without taking things apart, if we recall the voltages around the circuit equal the source voltage—always. If there is a bad connection, a too-large part of the supply (battery) voltage will appear across it under starting load. When we spot this we may have something useful to work on. The device we use here is a peak voltage reader.

Peak Voltage Reader

A bad-joint resistance of, say, one-tenth ohm is hard to measure. But at typical starter current levels of several tens to a few hundreds of amperes, voltage across it may be

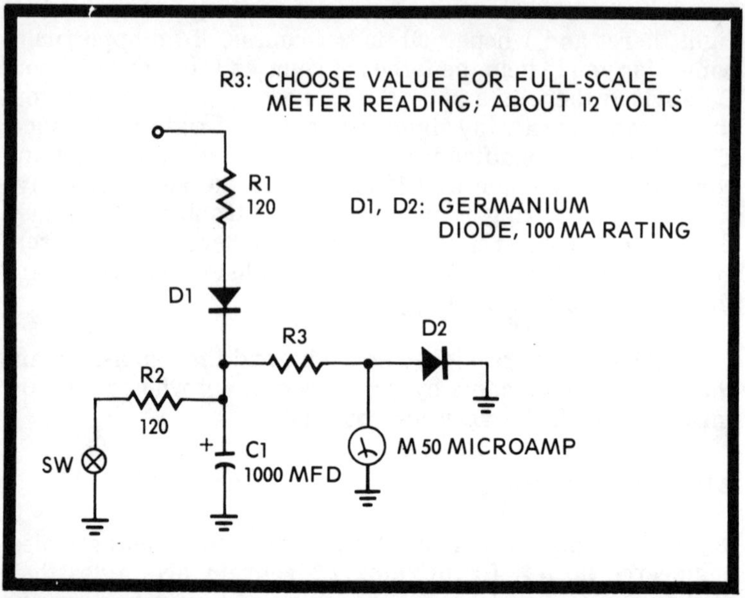

Fig. 9-20. This circuit remembers (briefly) the peak voltage applied at its input.

anything from one volt to practically the entire battery voltage. But this happens only under starter load. We want to try a start, and then see if, in the course of our attempt, much battery voltage is used up across a joint rather than across the starter terminals.

Closing switch SW in Fig. 9-20 shorts C1 through 120 ohms, discharging it after any previous test. Then Terminal T1 is attached to the positive side of the suspected bad joint or cable and Terminal T2 to its negative side. A few seconds of starter test develops some voltage across T1 and T2, charging C1 through R1. C11 develops a voltage at about a tenth-second time constant, which is comparable to the time scale of events in the starter circuit. As the test is finished, C1 begins discharging through R3 and the meter. R3 is large and the discharge slow; there is time to comfortably get up from the drivers seat or deep under the hood and look at the meter to see what happened. Since D1 is a germanium diode dropping about 0.2 volts from input to capacitor, any reading at all on meter M is reason to be suspicious of the joint or cable the instrument was observing. Diode D2 prevents excessive readings from damaging the meter.

Battery Checker

Now, perhaps, we wonder if the battery is holding up. A no-load voltage test may come out quite well even if the battery has truly failed. We have to see what the battery does when asked to do real work. Thevenin's theorem tells us a sick battery will have a large equivalent resistance, which we observe by seeing a large voltage drop under load. Evidently, we need a peak reader dual, a "trough" reading circuit. This is easy to make; a working sample appears in Fig. 9-21.

Closing switch SW charges C1 to the battery resting voltage. Then a starting test discharges C1 through D1 and current-limiting resistor R1 to the lowest voltage reached by the battery during the test. A large drop across the battery terminals (not the clamps connected to the terminals) indicates the battery is failing to deliver current, or else is overloaded. A small voltage drop is good news or else indicates little current is drawn by the starter circuit.

These two circuits might be combined, adding a switch to choose between one or the other, to make a useful automotive test instrument. A word of suggestion. Working around cars or any other machinery (even electronic) will prove this. It's not how elaborate or ingenious a test instrument is, it's what it tells you that counts. For many applications a lamp with two leads and clips on their ends will tell you as much as you need to know. That's simple—and it's cheap.

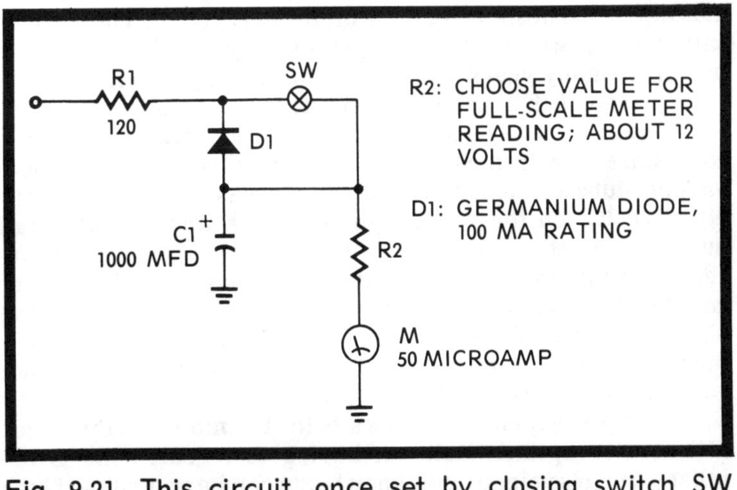

Fig. 9-21. This circuit, once set by closing switch SW briefly, remembers the **lowest** voltage applied across its input. It is a dual of the circuit in Fig. 9-20.

Chapter 10

Test Instruments

Once your lab is set up with the needed basics, you know something about how circuits work, and you have finished some projects, maybe you will become interested in new test equipment. This is good, but such an interest must be kept under control. Industrial and military surplus of the small, useful type is one way to keep costs down. Building your own is possible, desirable—and challenging. This is because test gear circuits verify how well you understand what you are doing in electronics. A successful hi-fi system, a good radio receiver, or some interesting control circuits—all are good experience. But when you have thought out and finished a few electronic circuits good enough for new circuit work, you've passed a milestone. Test gear work separates the men from the boys.

Another reason for making your test gear is that once you have mastered the technical side, you can design instruments better suited to your work than most you can buy ready-made. The manufacturer selling to the overall market must hit an average in his designs—a little bit for everybody. You want to satisfy exactly one customer, and if you do it well the result is useful for many years. And if you miss, you get to live with it until you fix it. Either builds confidence and good character.

Finally, you save money. This was not always true, but solid-state electronics works better than the historical vacuum-tube version. Now, with an increasing variety of very effective ICs on the market you can pick and choose small numbers of these to assemble into fine if slightly unconventional test gear. An example is any digital timer or counter. In the early 1960s the man who thought about building his own was out of touch or else some kind of a fanatic; now a modern counter from ICs is rather less of a challenge than a good shortwave receiver or a hi-fi rig.

A simple, expensive trap waits for the man who thinks too little or too optimistically in buying test gear. This is the fallacy that less electronics knowledge can be neutralized by a little more test gear investment. There is a small grain of truth in this, but much wrong. Events in electronic circuits proceed systematically (for those who understand the principles) and it isn't necessary to see the whole picture to know it. But if you haven't the basic understanding needed, the most graphic whole illustration won't bring comprehension. For example, read an engineering textbook. Accordingly, choose your test gear setup for the least you can get by with—and be inventive. The test gear list in Chapter 2 probably looks short, but it's realistic.

MAKING A STANDARD VOLTAGE

A classical electronics amateur's problem is finding an accurately known voltage for test and comparison work. This problem has a simple, satisfactory modern-electronics solution. The rather substantial terminal voltage variation of a set of common carbon-zinc D cells can be averaged out and preserved in an electronic circuit as a 1.54 volts standard reference source. And if you want, the source can be modified to deliver several additional known voltages, once the one is calibrated.

Fig. 10-1 is an appropriate circuit. D1 is a 6-volt zener biased through R1 to a current safely above its knee (which

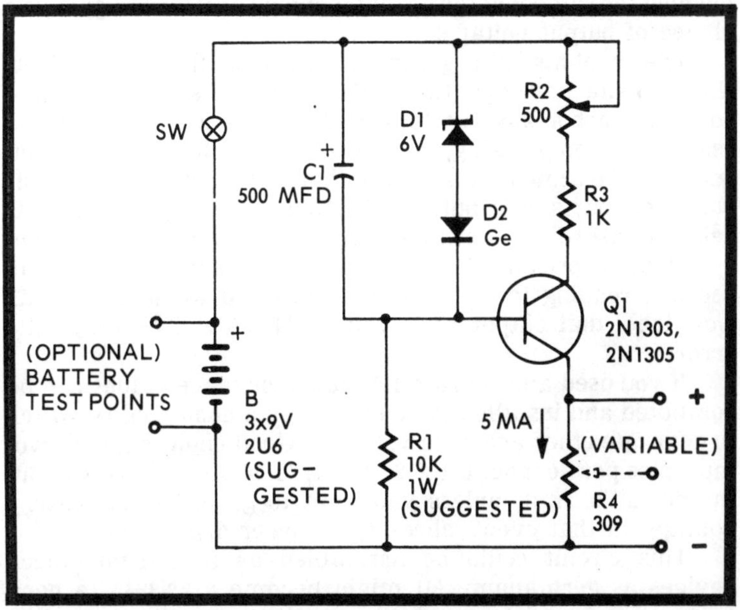

Fig. 10-1. This simple voltage standard is based upon a transistor circuit that generates a constant current.

should be located by test). A 6-volt zener typically has near zero temperature drift. Diode D2 adds to the zener's characteristics a temperature drift and a constant voltage similar to the diode drift and voltage of Q1's base-emitter diode.

If a fixed voltage is held across a resistance, a constant current flows through it (Ohm's Law). Q1 acts as an emitter follower holding the D1 voltage across calibrating resistor R2 and fixed resistor R3. The values shown are chosen for a current near 5 ma, which appears at the Q1 collector terminal, less the Q1 base current. The Q1 base current is roughly 1 percent of its collector current (beta equals 100, approximately), but this loss is constant and compensated by calibration. The several microamps collector leakage current is also unimportant for room-temperature operation. And the Q1 collector current is essentially unaffected by collector voltage, which doesn't vary much anyway. Accordingly, we can suppose the current in R4 is 5 ma, whether R4 is accidentally replaced by a short circuit, or if voltage across a larger R4 leaves only two volts or so (it should not be less) of battery voltage for Q1. This means that, once we have fixed the Q1 collector current accurately, a stable R4 will always develop the same voltage or that a choice of R4's will give us a choice of output voltages.

The 309 ohms value given for R4 is near the preferred 308 ohms (5 ma develops 1.54 volts). This sets us up for the nominal carbon-zinc battery voltage. Once the circuit is assembled and working, it must be calibrated. For this you need a large-scale multimeter with a 2- to 2.5-volt range. Visit your local parts distributor and find a box or three of fresh D cells. Measure the terminal voltage of every one of them (you'll be surprised at the variations). Average the results and use this reading to calibrate the standard by adjusting R2. Now label that output "1.54 volts." There will be very little error.

If you used an accurate R4, new resistance values can be computed and installed to be chosen by a good-quality switch to generate other accurate voltages. Good choices go by two-multiples per decade: 0.1, 0.2, 0.5; 1, (1.54), 2, 5. This is about the circuit's limit, unless you want to go to higher battery voltage. In that event, check Q1's power dissipation.

This circuit could be assembled as well from silicon devices as germanium. Q1 might become a 2N3944 (a good G.E. utility transistor) and D2 a small-signal silicon junction diode. The circuit would then become an NPN dual of that shown in Fig. 10-1.

For accurate work it's important to understand how this standard reacts to loading. The picture is simple: the load simply steals some current from R4, and it's usually easy to estimate how much. For instance, if a meter sensitivity is 20,000 ohms per volt and the 5-volt range is in use, meter resistance is 100,000 ohms. This resistance takes ten microamps per volt applied across it. When calibrating at 5 volts, 50 microamps is going to the meter rather than developing voltage across R4. Since R4 is about 1000 ohms the error is 50 millivolts, which is low; probably about the width of the meter needle. The finished calibrator is shown in Fig. 10-2.

A SIMPLE AC VOLTMETER CIRCUIT

Small audio voltages are often important, and any lab ought to have some way to measure these. A fairly expensive AC voltmeter usually sits around waiting for this work, but here is a circuit that adapts almost any multimeter to do the same job. Since it uses two uA709 operational amplifiers (or a pair of 741s) it is less simple than it seems, but this is a characteristic of op-amp circuits, a nice characteristic. As shown in Fig. 10-3 the circuit measures only 0-10 millivolts, but this basic sensitivity is easily reduced or increased.

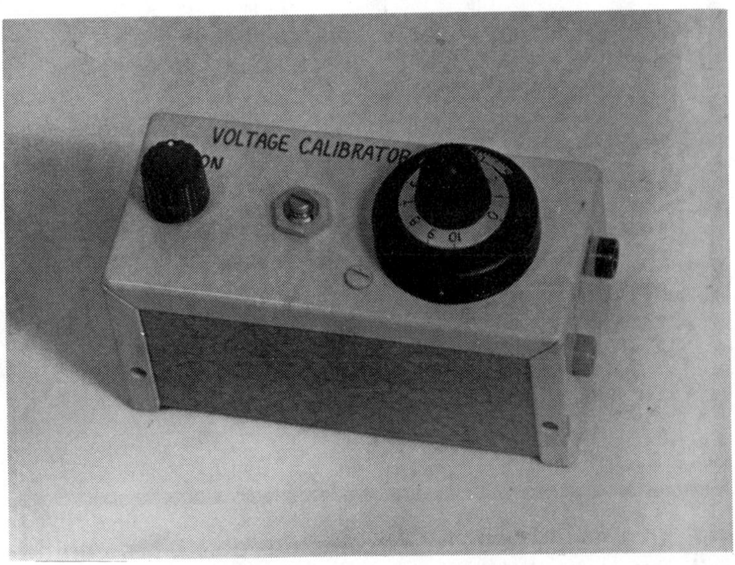

Fig. 10-2. The circuit of Fig. 10-1 assembled into a 10-volt accurate voltage source. Batteries are three 2U6 cells.

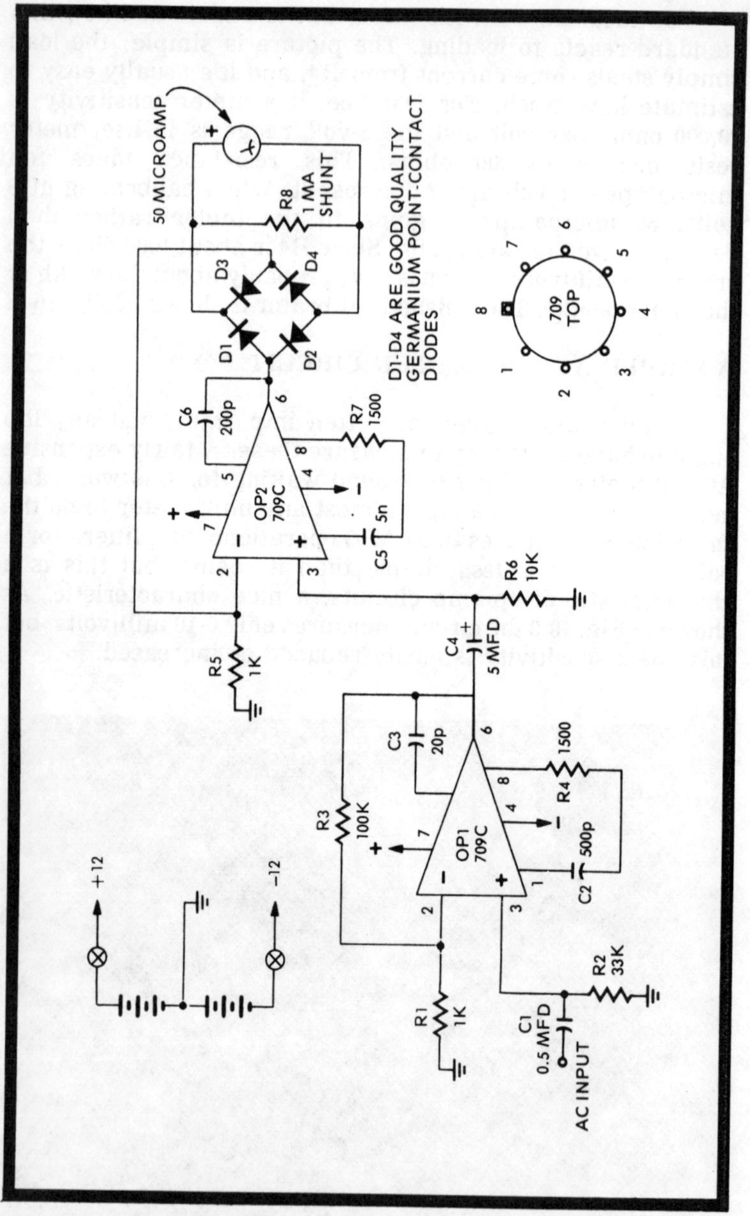

Fig. 10-3. This basic circuit combines a x100 preamplifier with a current-source meter driver to measure 0-10 millivolts AC. Basic sensitivity of the meter driver circuit is 0-1 volts AC.

OP1 is a x100 preamplifier to build up a 10 millivolt input to the 1 volt required by the OP2 meter driver. Its x100 voltage gain is instantly apparent to anybody familiar with operational amplifier circuits, since the ratio of R3 to R1 is 100. C3 and C2-R4 are frequency compensation for the 709, they're not needed if you build with the 741s. All high feedback circuits need compensation, even if the manufacturer has tucked it inside the case at a slight loss in utility. The 33K R5 fixes the input resistance at a level appropriate for most solid-state electronics work, but R5 has another job. It chooses which end of C4 is plus.

Which end is important, since the R6 end of C4 is near ground voltage and OP1 can deliver any fixed DC output from -10 to +10 volts. But we recall the maximum 0.5 microampere bias current needed by 709 type ICs, and since this is charge flowing **into** Terminal 3 (NPN input transistors), then Terminal 3 must be slightly negative with respect to ground. This reappears at the output as a fixed offset near -2 volts so that we now know the plus end of C4 faces R6. C4 ought to be a high-quality capacitor, tantalytic preferred. If you change the preamplifier gain, remember to refer to Fig. 7-12 to discover the appropriate new frequency compensation.

The OP2 meter driver circuit will look interesting until about the end of the next paragraph, when you will see how simple it all is. We begin with an inside detail: the diodes D1-D4. How do we get these diodes, which are naturally nonlinear at very low currents, to behave essentially linearly? Simple. We put them in an op-amp feedback loop as you find here. The op-amp forces them to behave nicely by feeding a slightly weird looking signal into them that corrects for their deficiencies. Now we look at our current feedback circuit.

Suppose one volt of AC appears at OP2's non-inverting input (pin 3). OP2 obeys the rule op-amps try for zero differential input, and so it returns an output that develops a signal at its inverting input equal in phase and amplitude to its input signal. But to do this, since there is feedback without a series resistor, it returns a current **into** R5 that develops the needed voltage **across** R5. This provides the indicated feedback, and if we are applying 1 volt AC at pin 3, then 1 milliamp AC must return to R5 to develop the 1 volt needed here.

Meter M is preferably a 50 microamp meter, purchased as a multimeter that can do other jobs around the lab. It is shunted to give it a 1-milliampere range. A 1-ma meter could work here, or a more sensitive meter or smaller R5 would increase OP2's circuit sensitivity. This lab circuit can be evolved into a useful instrument by adding a voltage divider at

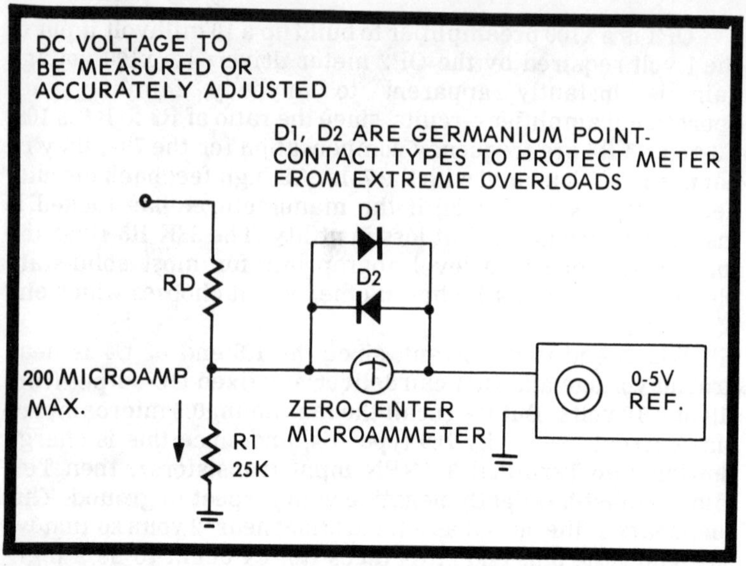

Fig. 10-4. This circuit measures an unknown voltage by comparing it with a known variable 0-5 volts. A voltage divider can be used to bring the unknown voltage into the proper range. Performance of this circuit is limited by meter sensitivity.

the OP1 input, OP2 input, or both. A good design probably would bypass OP1 for measuring voltages above 1 volt, since voltage dividers can be frequency-sensitive. Adding a pre-preamplifier for microvolt sensitivity is an interesting thought—and it will introduce you to the problem of inescapable natural circuit noise.

AN ACCURATE DC VOLTMETER CIRCUIT

Good voltmeters are typically expensive, though they contain simple circuits. Much of their price goes into the finely built meter movement—a part of the instrument we don't really need. The system described here doesn't quite eliminate the meter, but maybe you could find a way to do that. If you follow up the difference amplifier here with an op-amp feeding a transistor lamp driver, using a strong feedback loop around the whole, you might be able to use the lamp rather than a cheap meter to indicate nulls. Isn't it interesting that all this can be cheaper than a cheap meter?

We begin with a way to get an accurately known 0-5 volts. If you are thinking about the standard voltage circuit in Fig.

10-1, maybe you've guessed we might replace resistor R4 with a 5K helipot. Using a 10-turn dial at 100 divisions per turn will give us 5 millivolts per division at an overall accuracy near 0.1 percent, if we can get one accurate calibration to start. If we use the dry-cell averaging calibration method suggested earlier, the appropriate helipot setting is 308, not 154. With that finished, we are ready for a very simple comparison circuit, just to get started thinking. See Fig. 10-4.

Applying the unknown voltage sets up a 0-200 microamp current through voltage divider RD-R1, which you design for your own needs. This develops 0-5 volts across 25K resistor R1, an accurate unit or something you've built up. (A good wheatstone resistance bridge is nice for that.) Then the reference voltage is adjusted for zero meter reading, which means the voltage across R1 equals the reference voltage. This is the "null setting," the key to measurements of this kind. At null, zero current enters or leaves the RD-R1 voltage divider through the meter, and so loading on the unknown voltage is that of the divider alone. Also, no current is stolen from the reference source, so no correction is needed there.

Now let us suppose we are measuring the rather small voltage of 0.5 volts. We discover a weakness of this null system. Our null indicator isn't perfect: it shows an acceptable null over a range of perhaps 10 microamps either way. This gives us 5 percent or better accuracy near 5 volts, with 200 microamps flowing down the divider. But at 0.5 volts and 20 microamps accuracy is completely shot.

One remedy for this problem is to measure only relatively large voltages (an appropriately chosen divider). Another is a more sensitive null indicator—read carefully. Still another, an approach that was right out of sight for amateurs a few years ago, is to bring in two or three operational amplifiers, as shown in Fig. 10-5. The principles are the same as above, but this circuit gives very sharp null indications. In fact, the nulls are almost too sharp, and there is no simple remedy to this problem. We'll come back to that. For now, let's begin with the OP1 voltage follower.

The OP1 circuit is called a voltage follower because it copies the reference voltage at all values, taking no current from it. Since the OP1 output is connected directly to its inverting input, output follows non-inverting input with a maximum error below 1 millivolt. OP1 can deliver appreciable current without upsetting this accuracy, though it doesn't have to here. R1-C1 is a simple filter to eliminate possible contact noise from the reference source helipot. With this system's sharp null, any such noise is a nuisance. Notice

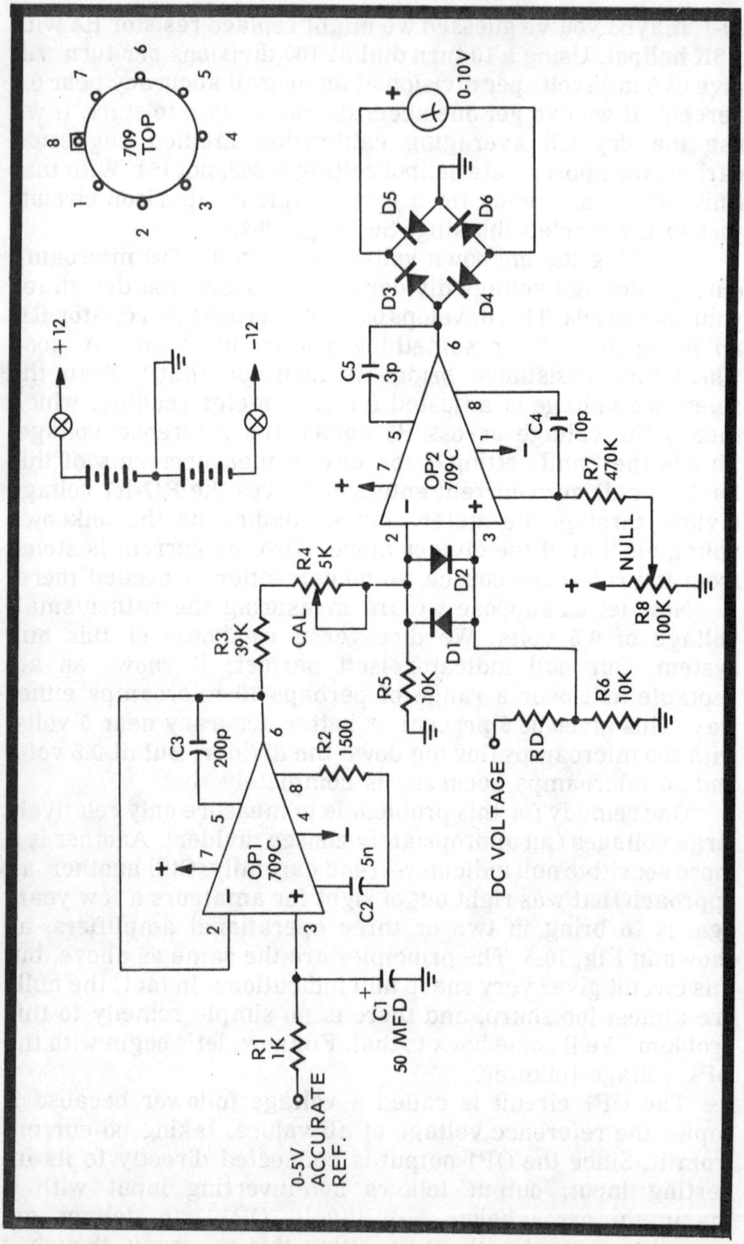

Fig. 10-5. A more elaborate comparison circuit uses two operational amplifiers. A third could be added to reduce loading on the unknown voltage.

the zero-db frequency compensation, which is not needed if you use a 741 here.

OP2 is a high-gain difference amplifier without feedback. Its inverting input sees 0-1 volts developed through R3-R4 and R5 from OP1, and its non-inverting input sees the voltage to be measured. RD represents a possible voltage divider system. When these two inputs are seen by OP2 to be the same, then the OP2 output voltage is zero. If they are different, the OP2 output is at least 15,000 times the difference, but not greater than 10 volts. This can be either polarity, therefore, diode bridge D3-D6 is provided so the meter always reads upscale. The first time this circuit is used, adjust nulling pot R8 for zero output at zero difference input. This adjustment is very sharp.

In normal testing, the input voltage to be measured is connected and then the reference voltage is advanced from zero (or wiggled around the expected value) while watching the meter closely. At a certain point, the needle will flick; a return to that point and careful adjustment will spot the null. This doesn't need to be a very accurate null. For instance, at a reading of 5 volts output the reference input is within 0.7 millivolts of the signal input (why?) and if that is 1 volt, then the accuracy is about one part in a thousand. This is better than the reference source can do, so the reference voltage accuracy sets the overall system accuracy. Once calibration of the source is achieved, performance should lie well within 1 percent.

Now, about OP2 feedback. Some correctly designed feedback could make the OP2 null much nicer to live with. There is none here because we are looking at a difference signal to be reduced to zero in the presence of a large common-mode signal. Feedback controls difference signal characteristics, but in this application we can have zero op-amp output at a 1-volt DC input. The normal op-amp rules don't apply here. So the only thing left to do is to put up with very sharp nulls, but they aren't that bad after all.

A SIMPLE LOGIC PULSE GENERATOR

Digital circuits lab work goes better if there is something on hand to replace parts of the system that are unfinished or located somewhere else. Industrial labs use versatile and expensive pulse generators for this, but if we restrict our interest to positive logic at most-used signal levels, and if we decide to get by without a wide range of frequencies, then it is easy to make a useful (and instructive) circuit.

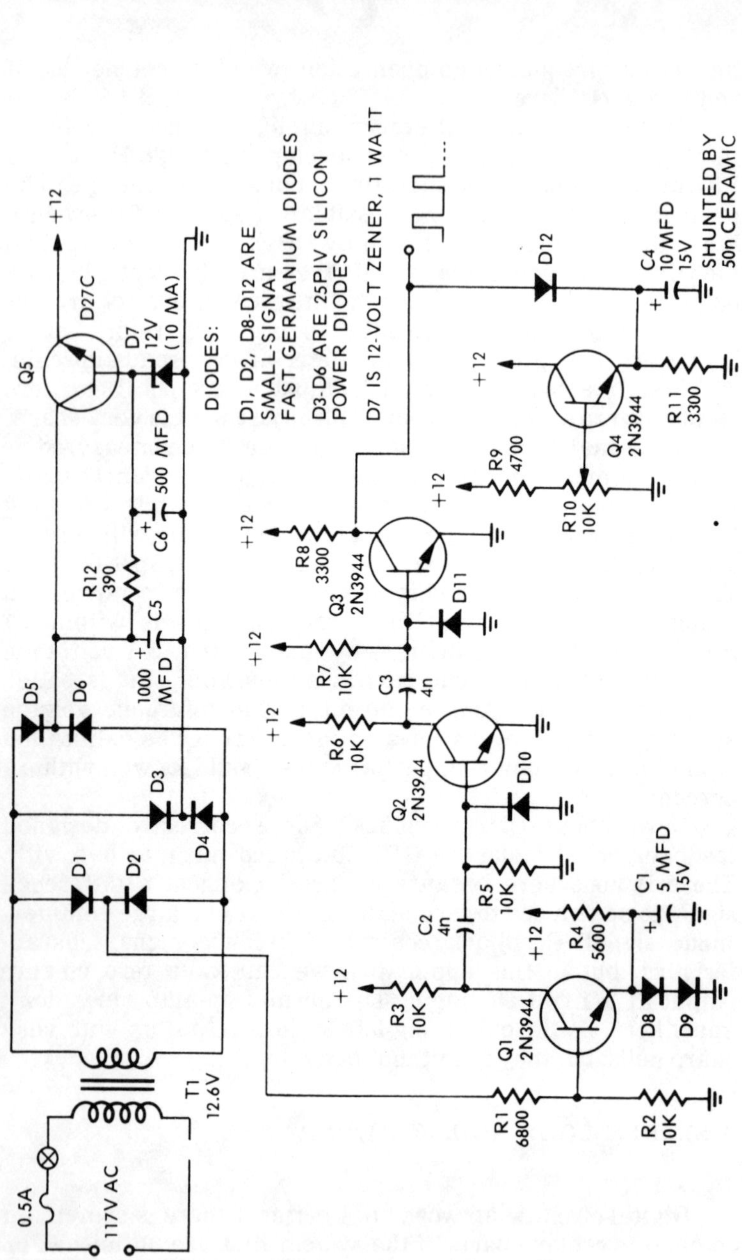

Fig. 10-6. A simple variable-amplitude pulse generator circuit. Parts values aren't critical and its output can be observed on service-variety scopes.

The circuit in Fig. 10-6 has an output that rests near ground level—a logical zero, but it makes regular, brief excursions to a logical 1. The repetition rate is fixed at 120 per second, and the logical 1s last about 40 microseconds. This duration is easily increased by using larger coupling capacitors in the circuit. A clipper limits pulse height without increasing output resistance or rise time, with height variable from about 0.5 volts to 6 volts. The output can be adapted to the nanosecond characteristics of digital ICs by adding a NAND gate breadboarded nearby, but its 100 nanosecond rise time and 200 nanosecond fall time are adequate for most discrete components work. All this—and it's simple.

A small transformer supplies 12 volts RMS AC (the original used two 6.3-volt transformers in series). Diodes D3 through D6 are a wheatstone bridge rectifier that feeds emitter-follower regulator Q5 for a fixed 12 volts output. The output is regulated because its value is used in fixing pulse height. Diodes D1 and D2 work with D3 and D4 as a second wheatstone bridge whose output is unfiltered. This gives us a "London-bridge" wave, as shown in Fig. 10-7. This little game with diodes, which might make the power supply puzzling, was carried out just to get the nice sharp corners in the London-bridge wave, and we now feed these corners to the base terminal of Q1.

But the Q1 emitter is up on some diodes, so that facing the rather large input signal (protective resistor R1 prevents damage) it spends a little time turned off and a lot of time hard on, with sharp transitions. The resulting positive-pulse output is fed through C1 to the Q2 base, with diode D9 present to catch

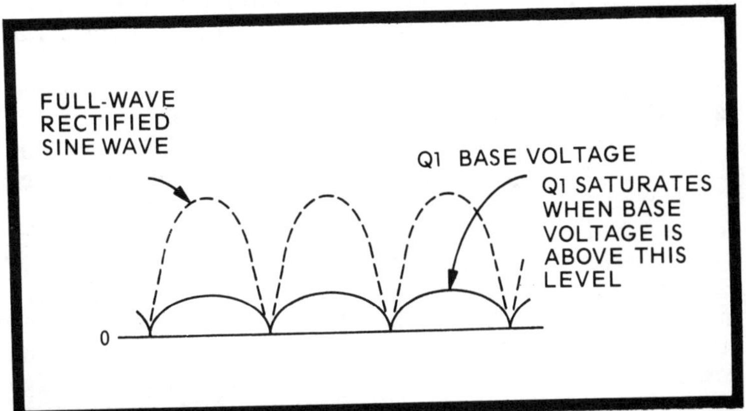

Fig. 10-7. The London-bridge wave which is applied to Q1's base.

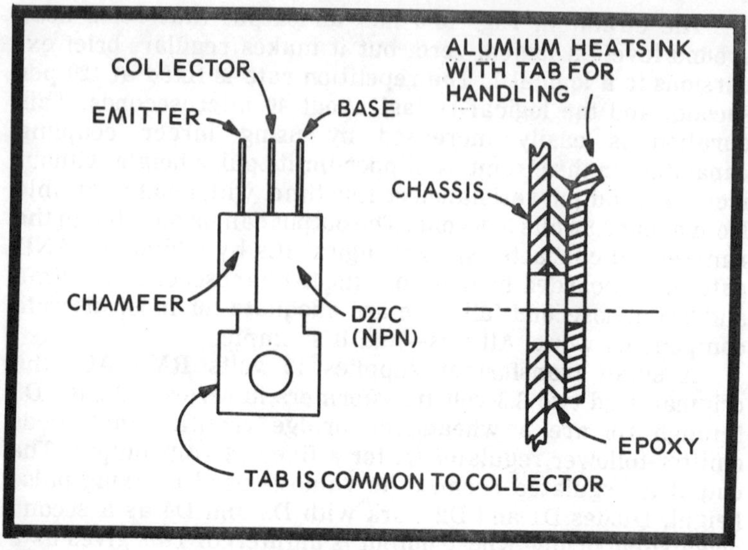

Fig. 10-8. These sketches show how to mount the D27C and similar transistors without any special hardware. The scrap aluminum and epoxy resin should be on hand in any busy lab.

any negative-going transitions. This prevents Q2 from getting turned off so hard it responds poorly to the pulse input. Q2 serves here primarily as a pulse polarity inverter.

The Q2 output is fed through C3 to the Q3 base, where it appears as a sharp, strong turnoff signal. Q3 is biased to conduct quite heavily except when turned off by this input, and diode D10 is provided as another negative-drift catcher. The final signal, a steady DC state of around +0.1 volts punctuated by a sharp positive excursion every 83 milliseconds, is direct-coupled to the output terminal.

A little thinking about this circuit reveals it's hard to hurt it. Shorting the output to ground doesn't upset anything, since it's shorted to ground through saturated Q3 most of the time anyway. And all the components are inexpensive types so that a real effort would be needed to do more than a dollar or two of damage by connecting a wrong voltage into it. A good try might destroy Q3, which can be replaced for 50 cents or so. The circuit passing through D11 is the "ceiling circuit," a diode catching or clipping circuit that absorbs any output voltage above that set through emitter-follower Q4 from potentiometer R10. If Q4 is holding the voltage across C4 (which absorbs the tops of clipped pulses) at +3 volts, then as the pulse from the Q3 collector rises above 3 volts, D11 goes into conduction and dumps the excess into C4. If R10 is set a volt higher, then the output can rise another volt before getting clipped. This is an excellent way to control pulse height, since circuit loading has little effect upon rise time. Instead, more of the energy that passes into C4 and to ground under no-load goes to the output terminal, so that this circuit's apparent output resistance is extremely small.

The D27C is a utility power transistor generally available for replacement work at prices near $1. It is a plastic-cased NPN type rated at 8 watts with its tab at 70 degrees C. It could have been hung in air in the original breadboard, but was heatsinked on principle. The circuit it feeds might draw more power someday, in a later version. Since the tab is connected to its collector terminal (probably the silicon transistor is soldered to it) the heatsink must be insulated from ground. This is easy to do, as shown in Fig. 10-8.

An aluminum plate of one to two square inches is epoxied to the chassis over a hole that provides clearance for the transistor mounting bolt. Epoxy that works into the bolt area is scraped off when half-hardened. An ohmmeter check then proves the plate is electrically insulated from the chassis by the thin film of epoxy. This mounting scheme requires no special parts and is equally good for all kinds of power transistor mounting.

If this logic pulse generator doesn't generate exactly the output you need, I'd suggest you build it anyway. If its 120 per second rate is excessive, a string of IC bistables can reduce it. Whatever pulse you need, large or small, you should be able to breadboard a circuit quickly that will make it from this generator's output.

A SIMPLE UTILITY AUDIO OSCILLATOR

Any electronics workbench needs an audio test oscillator, or at least the man who works at it should know how to make up something quick. Two convenient circuits are the wein bridge and the phase-shift oscillators. Both are "RC oscillators"; they get by without the inductances and tuned circuits commonly seen in radio frequency gear. This avoids using inductances in the henry range which are typically heavy, expensive, or both. Either circuit can be assembled without using special components.

The wein bridge is technically nicer than the phase-shift oscillator. It needs fewer frequency-fixing components and, using a simple amplitude-controlling system, it generates a

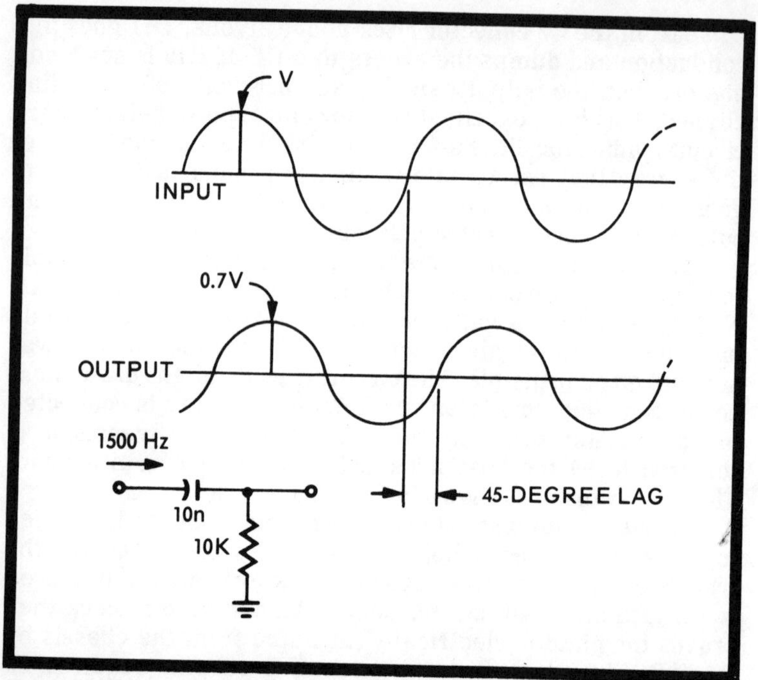

Fig. 10-9. The output voltage of this RC divider is slightly out of step with the input. The lag will be greater if another divider follows the first one.

nearly perfect sine wave. It offers better frequency stability than the phase-shift oscillator. Both the wein bridge and the phase-shift oscillator work well at frequencies under 1 Hz, but also at 500 kHz or better.

Because the phase-shift oscillator is simpler and easier to understand, we consider it here. It's a nice circuit for one-off test oscillator breadboarding, or for something built into an electronic system. It is easily assembled from inexpensive parts and not very sensitive to loose tolerances. It generates a passable sine wave, and for critical applications a small manual adjustment can be added to improve its output to very near perfection. To see how it works we begin with the RC network in Fig. 10-9.

This network is a frequency-dependent voltage divider, here we look into details of its performance. Suppose we apply a sine wave to this. Near the corner frequency, the output is not only less than the input, but is out of phase with it. We say the output "lags" because if we chart it (or look at it on a scope) we see it crosses the zero line a bit later in time than the

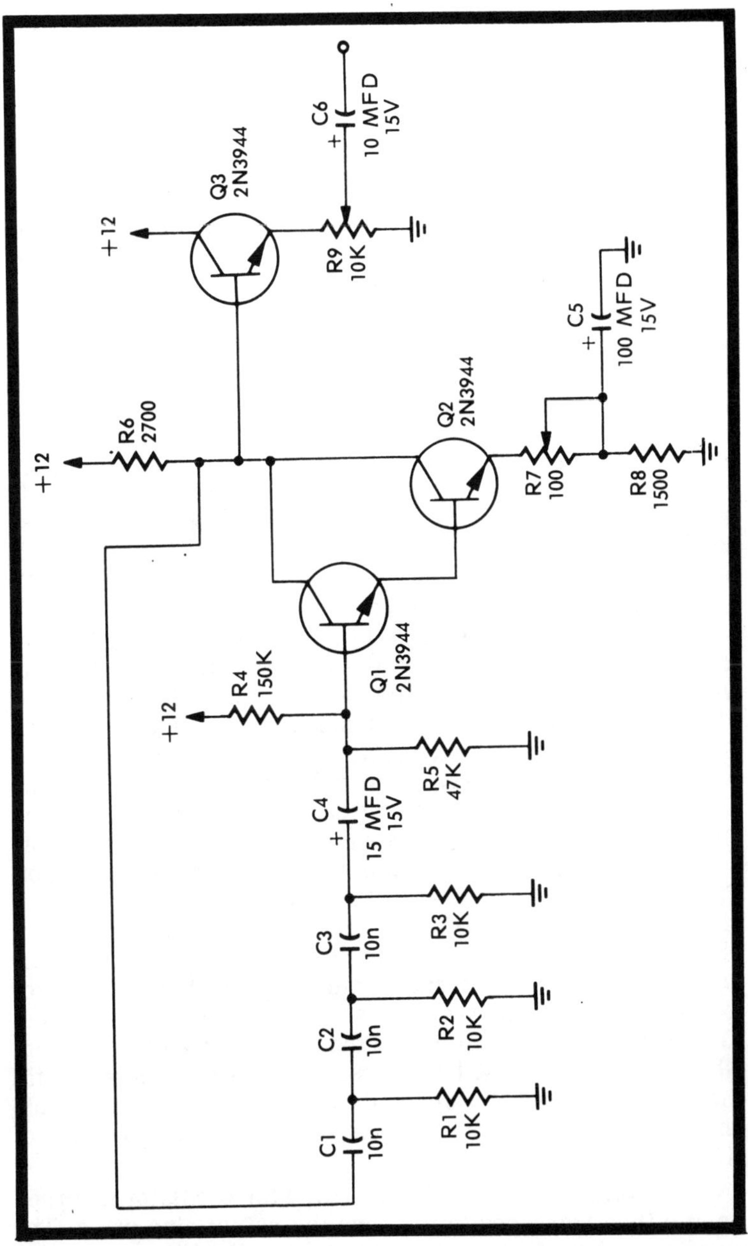

Fig. 10-10. Three RC voltage dividers in series can give the input signal a lag of 180 degrees. A practical application is this phase-shift oscillator.

input sine wave. Since the same condition that results in lag also reduces output we must be conservative about this—but why can't we put several of these RC circuits in series to get more lag? This is the basic idea of the phase-shift oscillator in Fig. 10-10.

Here, R1-C1, R2-C2, and R3-C3 are three phase-lag RC circuits in series. They provide a signal path from the common-emitter Darlington amplifier Q1-Q2 collector back to the base terminals. Now, if we suppose a sine-wave signal exists at the Q1-Q2 base, what goes on in the rest of the circuit? The signal appears at the Q1-Q2 collector terminal reversed in phase, which is the same as a 180-degree phase lag (a basic math book will help work this out). The signal is also about 30 times larger than the input, an important detail. The sine wave now goes through the three RC networks and is retarded by each in turn to acquire a further 180 degrees of lag, which brings it right back into step with the input. And its amplitude is reduced by 30 times. A clever choice of base circuit components C4 and R4-R5 could replace C3-R3, but this complicates explaining the circuit. It seems nicer to make C4 extremely large, and then leave it out of the picture.

Since the return signal is in step with the input, it reinforces the input; if it equals the input (which we supposed, without saying where it came from), it can replace the input. Then, this circuit will run on indefinitely, providing its own drive signal—which makes it an "oscillator." It works at a particular frequency because at only one frequency does the phase-shift arithmetic of 360 degrees lag around the loop work out right, and that's why it is called a phase-shift oscillator.

Signal amplitudes around this circuit are very important. The loss through the phase-shift network is fixed near a factor of 30. If the gain of Q1-Q2 exceeds the loss, oscillation amplitude will increase until something limits it; the transistor going into cutoff or saturation during part of the cycle, for example. But this causes distortion which we want to avoid. The answer to this is to adjust the transistor gain to a point where the circuit just barely oscillates. Then, there is still limiting action, but it is so gentle that distortion is negligible and cannot be seen on an oscilloscope. This is the purpose of R7.

R7 is adjusted to zero the first time the oscillator is turned on and then increased to improve the output sine wave. The original lab breadboard developed an excellent sine wave with R7 at about 33 ohms, which was very near the point where oscillation stopped entirely. Since the oscillator gain is trimmed to a rather sensitive point, it's good practice to use emitter-follower Q3 to minimize loading. If its output resistance through R9 is excessive, a second emitter-follower can be added to get a variable amplitude low-resistance output.

A THIN-LINE PULSE GENERATOR

Many shortwave receivers, particularly the less expensive types, tend toward a "relaxed" calibration accuracy. There is a simple, commonly used solution to this. A stable, accurately calibrated oscillator physically near the receiver, or perhaps built into it, provides marker signals at precisely known frequencies. The receiver operator then applies the appropriate English to the dial calibrations near the marked frequencies. The more technically oriented operator may spend time trimming the receiver tuning characteristics to fit the manufacturer's calibration scales, or he may use a marker oscillator to calibrate a receiver he has just built.

Marker generating circuits are simpler than many people think. Basically, they are often crystal oscillators operating at 100, 200, or 1,000 kHz. Sometimes these are provided with bistable count-down circuits to generate square waves at 10, 25, or 50 kHz, or at other frequencies. It is commonly believed these square-wave outputs are rich in odd and even harmonics, at all possible multiples of the fundamental frequency. This is false. See Fig. 10-11.

Here we find that a square wave's fundamental frequency is extremely powerful, as suggested by the large areas under the square wave. Low-multiple harmonics are also strong, but

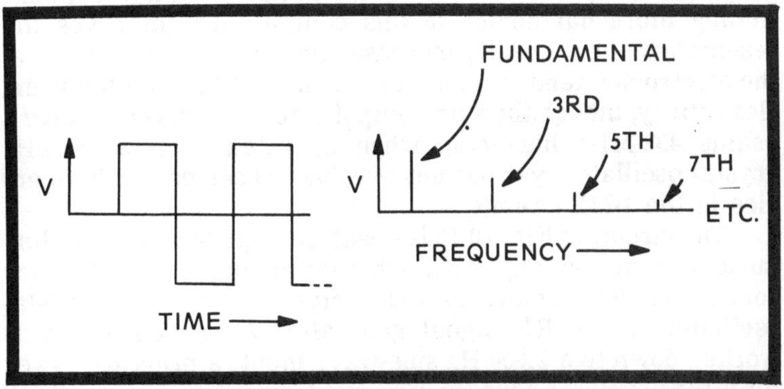

Fig. 10-11. The familiar square wave has a very limited frequency content. Only odd harmonics appear, and they are weak compared to overall signal strength.

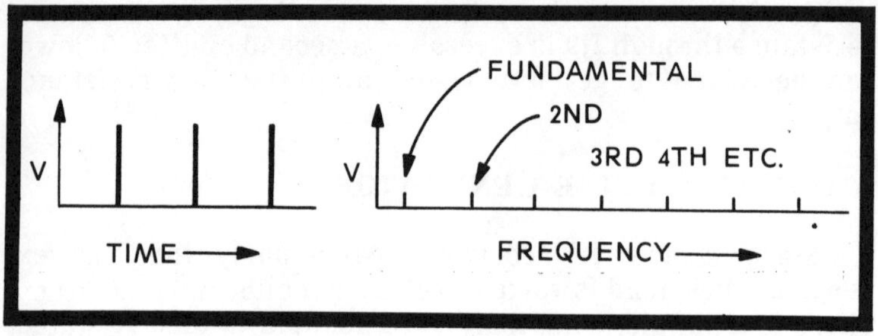

Fig. 10-12. A thin-line pulse contains a complete spectrum of odd and even harmonics up to very high frequencies.

the strength of harmonics falls off rapidly as frequency rises. And we discover these aren't any even-numbered harmonics at all. This result is based upon mathematics proven right. When a square-wave generator seems to put out even-numbered harmonics—there is something wrong with it!

If the square wave is wrong for marker generators, then what is right? It is a signal many amateurs never heard of, the thin-line pulse. See Fig. 10-12. This pulse encloses little space, so that its low-frequency harmonic content is weak. But the sharp rising and falling edges suggest lots of high-frequency harmonics. It turns out, as shown, the thin-line pulse has a harmonic spectrum consisting of odd and even harmonics, all about equally strong, up to very high frequencies.

A simple graphical demonstration of this is to chart a fundamental and four or five harmonics of it, all of equal amplitude, all starting in phase. The result, developed by graphical addition, looks something like a thin-line pulse. Adding more harmonics to this computation improves the resemblance, but greatly increases the work. It's easier to let the electronics tend to that. An instrument like the following (less utility, though the same output circuits) proved to have a usable 432-MHz harmonic when operated from a 100-kHz crystal oscillator, yet did not overload receivers at frequencies in the 1-MHz range.

The circuit in Fig. 10-13 is designed to generate a thin-line pulse from almost any input, whether an audio oscillator (for rough 5- or 10-kHz markers and interpolation work), a crystal oscillator, or an RF signal generator. A breadboard of it worked down to a 2,000-Hz sine-wave input, a practical lower limit for harmonic sources. The input signal can be anything (within reason) above 200 mv, which sees a load of about 10K. This signal goes to Q1, which works linear or overdriven, and proceeds to the first clipping and shaping stage, Q2.

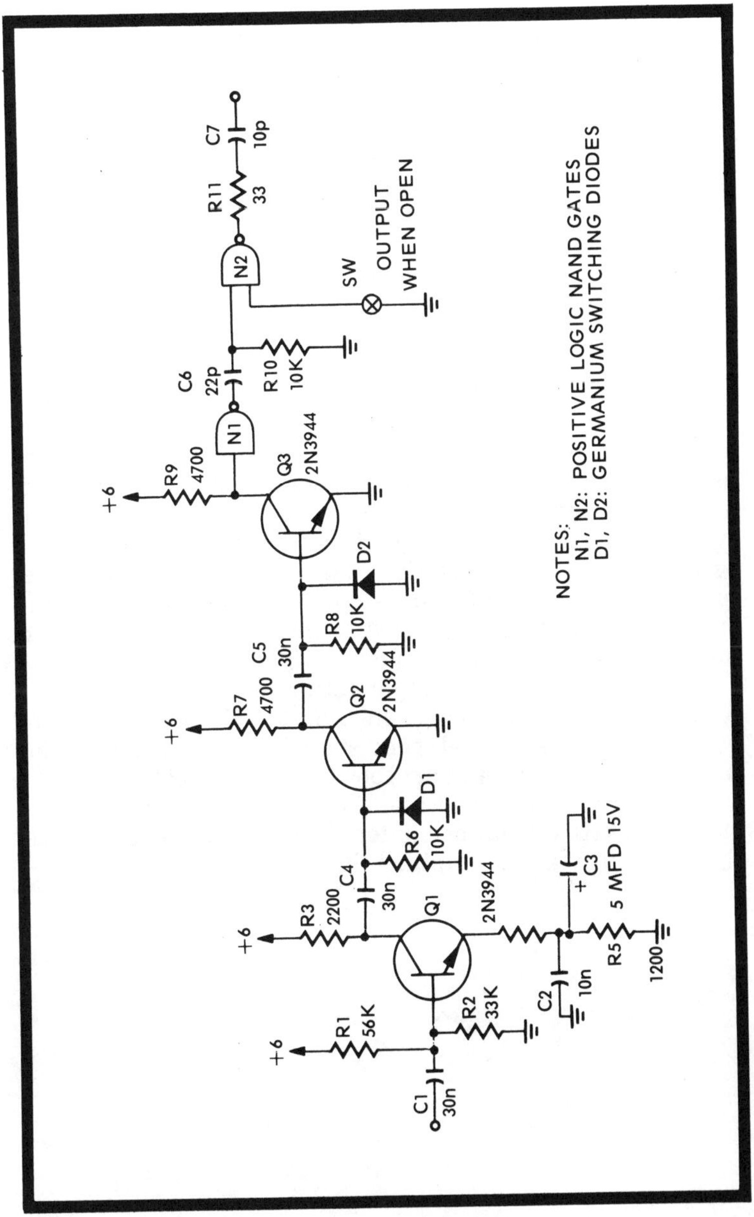

Fig. 10-13. A pair of fast IC NAND gates can turn a square wave into an excellent thin-line pulse by logic switching through a very small capacitor, C6. The IC supply lines must be well bypassed for the high frequencies generated.

Q3 further sharpens and amplifies the signal in preparation for driving the first NAND gate. This output drives the second NAND gate, now through a very small capacitor. The final output, appearing through a 33-ohm protective resistor, is a 5-volt 100 nanosecond pulse with rise and fall times of ten nanoseconds or better. This rather remarkable output is clearly visible on an expensive laboratory scope (Tektronix 545A) but may not appear at all on an inexpensive hobby scope. Its presence can be established by picking up its harmonics on a radio receiver. Repetition rate is the same as the input frequency.

Since the RF signal output sounds just like any other RF signal there needs to be some way to mark it. A simple solution is to use one of the NAND gate inputs as a logic drive fed by a toggle switch or a low-frequency astable oscillator. To get best results from this circuit, we must remember Kirchoff's loop rule. Our thin-line output is remarkably rich in high frequencies. These will appear at the output terminal and inside receiver circuits only if they aren't wasted across parasitic reactances. For this reason the NAND gate power supply line must be well bypassed right at the gate. And the IC ground return must be carried to the output ground return, together with good high-frequency signal wiring.

The thin-line signal output terminals may be a pair of banana jacks, a BNC connector, or both. Since this circuit generates a strong signal, good shielding may be needed, but that is mainly a matter of appropriate mechanical design. The thin-line generator would make a nice companion unit to a stable 100-kHz crystal oscillator, if both were built in boxes like the utility audio amplifier in Chapter 8.

Appendix

NUMBERS IN ELECTRONICS

The values in Fig. A-1 are the 12th and 24th roots of ten, if you are wondering where they came from. To see how ranges overlap, add and subtract appropriate tolerance amounts from the nominal values. It turns out the overlaps are small and consistent; all values are possible but none doubly emphasized. To save yourself a lot of trouble (and be much more effective than the next fellow who needs two minutes to decode a resistor's markings) memorize the 10 percent magic numbers and the color code. It's a small effort—and worth it.

10 PERCENT VALUES		5 PERCENT VALUES	
10	33	10	33
		11	36
12	39	12	39
		13	43
15	47	15	47
		16	51
18	56	18	56
		20	62
22	68	22	68
		24	75
27	82	27	82
		30	91

Fig. A-1. The standard 10 percent and 5 percent electronics components values. Look at your computation and at how the circuit works in choosing values in design work.

STANDARD NOTATION

To multiply two numbers in standard notation, first reduce both to standard notation. Add exponents, and write the result. Then multiply significant figures, and maybe you

$2500 = 2.5 \times 10^3$

$4{,}730.000 = 4.73 \times 10^6$

$.003 = 3 \times 10^{-3}$

$.00000086 = 8.6 \times 10^{-7}$

$330{,}000 = 3.3 \times 10^5$

$.030 = 3 \times 10^{-2}$

Fig. A-2. Examples showing how large and small numbers are rewritten into standard notation.

will want to adjust the answer for readability or convenience. See Fig. A-3. To divide two numbers, change to standard notation and write them out as in Fig. A-4. Maybe you will want to shuffle exponents a bit. Next, bring up the ten power from the bottom, reverse its sign, and add it to the ten power at the top. Then complete the computation by working out the division of the significant figures. To do a combined operation, work out exponents above and below the fraction bar, and rewrite as shown in Fig. A-5. Finish working out the ten powers and then do the significant figures.

$120{,}000 \times 6 \quad 1.2 \times 10^5 \times 6 = 7.2 \times 10^5$

65 microamps $\times\ 120 = 6.5 \times 10^{-5} \times 1.2 \times 10^2 = 7.8 \times 10^{-3}$ amps

100 kHz $\times\ 4320 = 10^5 \times 4.32 \times 10^3 = 4.32 \times 10^8$ = 432 MHz

Fig. A-3. Multiplication of large numbers is greatly simplified after conversion into standard notation.

$$\frac{150{,}000}{6} = \frac{15 \times 10^4}{6} = 2.5 \times 10^4$$

$$\frac{1}{382{,}000} = \frac{1}{.382} \times \frac{1}{10^6} = \frac{1}{.382} \times 10^{-6} = 2.62 \times 10^{-6}$$

Fig. A-4. Standard notation simplifies division, too.

$$\frac{330{,}000}{1500} = \frac{3.3 \times 10^5}{1.5 \times 10^3} = 2.2 \times 10^{5-3} = 2.2 \times 10^2$$

Fig. A-5. A combined operation. Remember to change the sign of the lower ten power when bringing it above the fraction bar.

Index